CONSTRUCTED ECOLOGIES

Today, designers are shifting the practice of landscape architecture towards the need for a more complex understanding of ecological science. *Constructed Ecologies* presents ecology as critical theory for design, and provides major ideas for design that are supported with solid and imaginative science.

In the questioning narrative of *Constructed Ecologies*, the author discards many old and tired theories in landscape architecture. With detailed documentation, she casts off the savannah theory, critiques the search for universals, reveals the needed role of designers in large-scale agriculture, abandons the overlay technique of McHarg, and introduces the ecological and urban health urgency of public night lighting.

Margaret Grose presents wide-ranging new approaches and shows the importance of learning from science for design, of going beyond assumptions, of working in multiple rather than single issues, of disrupting linear design thinking, and of dealing with data. This book is written with a clear voice by an ecologist and landscape architect who has led design students into loving ecological science for the support it gives design.

Margaret Grose teaches landscape architecture in the Melbourne School of Design, within the University of Melbourne, Australia. Rarely for a trained landscape architect, she also has a long earlier history as an agricultural scientist and ecologist, working in Western Australia as well as in Oxford and Cambridge in the UK, where she did experimental and theoretical research in mathematical biology. She has published more than forty-five journal articles and book chapters across biological science and design, and is an Associate Editor of Oxford's *Journal of Urban Ecology*.

CONSTRUCTED ECOLOGIES

Critical Reflections
on Ecology with Design

Margaret Grose

LONDON AND NEW YORK

First published 2017
by Routledge
2 Park Square, Milton Park, Abingdon, Oxon OX14 4RN

and by Routledge
711 Third Avenue, New York, NY 10017

Routledge is an imprint of the Taylor & Francis Group, an informa business

© 2017 Margaret Grose

The right of Margaret Grose to be identified as author of this work has been asserted by her in accordance with sections 77 and 78 of the Copyright, Designs and Patents Act 1988.

All rights reserved. No part of this book may be reprinted or reproduced or utilised in any form or by any electronic, mechanical, or other means, now known or hereafter invented, including photocopying and recording, or in any information storage or retrieval system, without permission in writing from the publishers.

Trademark notice: Product or corporate names may be trademarks or registered trademarks, and are used only for identification and explanation without intent to infringe.

British Library Cataloguing-in-Publication Data
A catalogue record for this book is available from the British Library

Library of Congress Cataloging-in-Publication Data
Names: Grose, Margaret, author.
Title: Constructed ecologies : critical reflections on ecology with design / Margaret Grose.
Description: Abingdon, Oxon ; New York, NY : Routledge, 2017. | Includes bibliographical references and index.
Identifiers: LCCN 2016037516 | ISBN 9781138890213 (hardback : alk. paper) | ISBN 9781138890220 (pbk. : alk. paper) | ISBN 9781315712543 (ebook)
Subjects: LCSH: Landscape architecture--Environmental aspects. | Landscape ecology.
Classification: LCC SB472 .G76 2017 | DDC 712--dc23
LC record available at https://lccn.loc.gov/2016037516

ISBN: 978-1-138-89021-3 (hbk)
ISBN: 978-1-138-89022-0 (pbk)
ISBN: 978-1-315-71254-3 (ebk)

Typeset in Baskerville
by Saxon Graphics Ltd, Derby
Printed by Ashford Colour Press Ltd.

Therefore again and again the earth deserves the name of mother which she has gained, since of herself she created the human race, and produced almost at a fixed time every animal that ranges wild everywhere over the great mountains, and the birds of the air at the same time in all their varied forms.

But because she must have some limit to her bearing, she ceased, like a woman worn out by old age. For time changes the nature of the whole world, and one state of things must pass into another, and nothing remains as it was: all things move, all are changed by nature and compelled to alter. For one thing crumbles and grows faint and weak with age, another grows up and comes forth from contempt. So therefore time changes the nature of the whole world, and one state of the earth gives place to another, so that what she bore she cannot, but can bear what she did not bear before.

Lucretius, *De rerum natura*, 5, lines 821–836
First century BC

Dedicated to the memory of my maternal grandfather
Neville Adolph Young (Jung)
who taught me about the bush

CONTENTS

Preface xi
Acknowledgements xv

PART I
A background to design 1

1 The environment is not a human construct 3
2 Global differences, not universals 14
3 Shifting adaptabilities, not static concepts 53

PART II
Thinking about design 93

4 Multiple, not solo voices 95
5 Inquiries, not assumptions 125
6 Thinking backwards, not forwards as a
   linear narrative 155

Concluding comments 185

Index 187

PREFACE

Time changes the nature of the whole world, and designers interact with that world continuously by making and constructing. How we interact with the world and how we design very much depends on what we think and how we think. This book contains critical reflections on landscape and design with a particular emphasis on ecology within design, with science as a fundamental cultural project in a world of change and fluxing conditions, places, and futures. Since adding landscape architecture to my existing scientific training in soils and plants, I have been continually struck by the incorrect and misleading ideas many people in the humanities have of science, such as believing that science is solely empirical, that science's ambition is about power over nature, or that science does not create. Indeed, science is about inquiry, about understanding how the world works, and the inter-relationships in that world. This hard-won effort of Western culture for more than 2,500 years, along with Judeo-Christian benevolence, laid the scientific foundations of clear questions, excellence, invention,[1] and creative freedom. This massive and expressive inquiry continues with such success that the rest of the world has followed scientific inquiry. While scientific knowledge fluxes, it gives insights into ways the physical, biological, and chemical world performs. Science is more nuanced than understood by those who consider it purely fact-finding positivism. Increasingly, science is concerned with uncertainty – not in correcting or eliminating uncertainty, but in managing uncertainty. This concern is important for designers because in the face of uncertainty and incomplete knowledge, we must make and construct.

One of the great challenges of science has been the relationship of evidence to theory. Here in this book, I have attempted to join evidence and theoretical ideas within ecological science to reflect upon design. This book has grown out of the challenge of teaching design students to understand that study need not and should not be bound to project examination or be precedent-driven – that theoretical ideas from science can strongly inform design. The question – how can ecological science inform designing landscapes? – drives the discussions and has come from my education in both science and design. Working across two disciplines means that I might be less likely to satisfy the scholars of either design or ecology.

I believe that the current lack of foundational knowledge within landscape architecture has led to a crisis in the profession. A lack of foundational knowledge is based on the pillars of how we are taught. The landscape historian is rarely trained as a historian; the person leading the ecological design course is usually not an ecologist; the design theoretician is usually neither a philosopher nor a historian. Thus, we build a profession on pillars that have shallower, more general foundations than are usually desirable for a strong structure – the engineer must know mathematics, the classicist must know Greek or Latin or both, because knowledge of substance allows critique of all that springs forth in a discipline. The absence of foundation is beginning to show in landscape architecture as we need to offer more substance to our interdisciplinary colleagues and to future generations of design students. The discipline of landscape architecture has, in its teaching, become a bit like teachers who only have a teaching degree, and not a degree in a discipline such as maths, literature, a particular foreign language, physics, chemistry, or biology. This has led to the odd question that the profession often poses to itself: what is landscape architecture? What do we do? Cardiologists do not ask the same question; more closely, rarely do architects.

Foundational knowledge in landscape architecture has been moving from horticulture and planting design to more performative testing of ecological systems. Yet dealing with ecological issues and ideas remains difficult for designers because it demands more informed forays into science. It is my belief that more substantial knowledge and more substantially based ideas are needed in landscape architecture. To work in shifting ecologies, shifting cultures, and shifting disciplines – all situations we face today – designers require detail, substance, rigour, tested knowledge, and exploratory ideas. For the constructed ecologies in much of current design practice, that means designers need to work with science in many of its forms.

I also need to give a word of caution. As design moves to greater use of scientific terms and takes on the notion of experimentation, this can be seen as dabbling if unsupported by good procedure and statistical advice, or can consist of 'experiments' that do not address the questions that the designers think they are addressing.

Most books in the design realm that contain words derived from the root 'construct' are in some way concerned with construction rather than theoretical ideas. In this book I am not concerned with the technical details of how a construction is designed and made to function, for whatever purpose it was designed, because technical details will change greatly over time and with variations in sites. Nor am I interested in restoration and its needs, or environmental management. Rather, I am interested in the thinking behind designs that have an ecological ambition – the ideas behind the constructed ecologies, and the ideas that have the potential to change the manner in which we construct ecologies in the future. This book is not 'theory to elaborate rules and procedures for production', considered by the landscape architect and theorist James Corner as a 'tyranny' of contemporary theory in landscape architecture.[2] This book is very much *not* a book about production of constructed designs but it does offer the tools to question the bases of our designs. My ambition is to encourage freedom from binds of thought and preconceived ideas, and freedom from some paradigms in current landscape design that I believe constrain design creativity.

In outlining ways of thinking about how we construct ecologies today in landscape architectural design, I work with the ways of 'knowing of things anew' of hermeneutic theory.[3] James Corner has defined hermeneutics as the subjective and situated construction of meaning; Richard Weller termed it a 'quantum and poetic view of reality'.[4]

Landscape architectural design is often defined by references to famous constructed ecologies such as the controlled French Baroque gardens in Versailles, the equally controlled elimination of the French Baroque by Capability Brown and Humphrey Repton and their creation of the English landscape and the picturesque in the mid-eighteenth century, the native plants of the prairie ideas of Jens Jensen, and Frederick Law Olmsted's Emerald Necklace in Boston. The nature writing of John Muir and others in the nineteenth century and the ecological movement of the twentieth have influenced design. In the last sixty years, there have been major additions to design thinking and process, inspired by both design and ecological science – Ian McHarg's *Design with Nature* in 1969, the articulation of systems theory, the now superseded ideas of succession and climax in vegetation from the 1930s,[5] the spatial ideas of landscape ecology that has grown into a field in itself, ideas of landscape urbanism, the comparative ecologies of cities (largely undertaken by ecologists), tensions in discussing native and exotic flora, tensions between site data to obtain generalised models and theories with design's specific groundings in site,[6] the problems of suburbia and the issues of multi-disciplinary design. Also, design has shifted away from the invisibility of natural processes (e.g. water put underground) to visible processes, shifted away from resistance engineering to permeable boundaries and flux, shifted from thinking in terms of a stable nature and a destabilising humanity to working with an unstable and changing

PREFACE

nature. It is a breathless mix and, as Richard Weller pointed out, the design professions have also been flooded by theory and ideas from architecture and philosophy,[7] with views across to architecture's new strivings between art, mathematics, physics, and engineering. More recently, performative design has risen to test and articulate the ecologies that we are constructing. While all of these influences shadow this book, the focus is ecological science in relation to design, and where our thoughts might go – shifting, fluxing, moving centres, testing sharpness over wide ranges, neither shutting in nor shutting out.[8] Because of these ambitions, the book is wide in scope, and all errors, of which I am sure there will be many, are mine.

Margaret Grose
Melbourne
March 2016

Notes

1. See Robert Friedel, 2007, *A Culture of Improvement: Technology and the Western Millennium*, Cambridge, MA: MIT Press. This book discusses the veins of invention that run through Western culture.
2. James Corner, 2014, 'Three tyrannies of contemporary theory', in James Corner and Alison Bick Hirsch (eds), *The Landscape Imagination: Collected Essays of James Corner 1990–2010*, New York: Princeton Architectural Press, p. 87.
3. Ibid. On p. 99 Corner discusses the ideas behind hermeneutic theory.
4. Richard Weller, 2014, 'Wordscape: The writings of James Corner in theory and practice', Afterword, in Corner and Hirsch, *The Landscape Imagination*, pp. 351–361.
5. Clement's theory of succession and climax is still used helpfully in restoration.
6. Elizabeth Meyer, 1997, 'The expanded field of landscape architecture', in George E. Thompson and Frederick R. Steiner (eds), *Ecological Design and Planning*, New York: John Wiley & Co., pp. 45–79. Meyer also discusses this in 'The expanded field of landscape architecture', in Simon Swaffield (ed.), 1997, *Theory in Landscape Architecture: A Reader*, Philadelphia, PA: University of Pennsylvania Press, pp. 167–170. I discuss the tension between the general and specific in Margaret Grose, 2014, 'Gaps and futures in working between ecology and design for constructed ecologies', *Landscape and Urban Planning* 132: 69–78.
7. Weller, 2014.
8. Inspired by *Corson's Inlet*, 1965 poem by the American poet A.R. Ammons (1926–2001).

ACKNOWLEDGEMENTS

This book has developed from lectures given primarily to landscape architecture students in the Melbourne School of Design within the University of Melbourne. To them I am grateful for many questions about design and ecology. They were often surprised to be taught genetics when they were expecting plant species lists, and to be given theoretical and ethical questions when they were expecting simple narratives, precedents in design, or textbook ecology. They even survived the odd physics equation. I would not have pursued the ideas of Chapter 6, which focuses on how to approach design, if I had not seen very positive changes in outcomes in design studio when students were liberated from standard site analysis.

Some influences run deep and I thank my best teachers whose lessons lie across the foundations of this book: Lex Parker, Alan Robson AM, and Richard Weller. I am indebted to fellow landscape architecture colleague at the University of Melbourne Jillian Walliss for her continual prodding and probing of my ideas, with many stimulating discussions about design, performance, and the need for theoretical thinking in landscape architecture. I thank the following people for their advice and expertise: Adam Carey and Barry Clarke, and other members of the International Dark-Sky Association of Victoria, Australia; Peter Hall FRS (dec. 2016), University of Melbourne, for discussion on inverse problems in mathematics and design; Fiona Johnson, University of Melbourne; Charles Massy OAM, Australian National University; Fraser Mitchell, Trinity College, Dublin; Alex Piel and Fiona Stewart, Ugalla Primate Project in Tanzania, Liverpool John Moores University and the University of Cambridge; and Deano Stynder, University of Cape Town. I thank my copy-editor Susan Schmidt, PhD, developmental editor of Beaufort Writing Group, North Carolina for great advice and for improving the clarity of my writing, and Sade Lee, editor at Routledge, for production advice.

I thank the University of Melbourne for a sabbatical in 2016 that allowed me to complete this book. At the MIT Center for Advanced

ACKNOWLEDGEMENTS

Urbanism, I thank Alan Berger and Celina Balderas-Guzman for time as a Visiting Scholar. In Massachusetts, I also thank the extraordinary generosity of Maseeh Hall at MIT, where I am indebted to Housemaster Jack Carroll. Finally, thanks to my children Markela and Kon for their long support, stimulation, over-the-shoulder criticism, and encouragement.

I thank the following organisations, offices, individuals, or printing houses for allowing images reproduced here: Alan Travers and BuroHappold Engineering and the Arriyadh Development Authority; Diana Balmori at Balmori Associates; Ryan Danby, Queen's University; Michel Desvigne, Paysagiste; Stephan Lautenschlager, University of Bristol; Fadi Masoud, MIT; Jeanine McDonald, Mararoa Appaloosas, New Zealand; Karen Morrissey OM, Meeline Station; Nelson Byrd Woltz Landscape Architects; Alex Piel and Fiona Stewart, Ugalla Primate Project; Daan Roosengaarde; Christie Stewart, and photographer Paul Verity.

Part I

A BACKGROUND TO DESIGN

1
THE ENVIRONMENT IS NOT A HUMAN CONSTRUCT

> There is sometimes an instant of delirium when a sensitive clavichord imagines that it is the only clavichord that exists and that it alone produces all the harmonies of the universe.
>
> Diderot[1]

The environment is not a human construct. At first glance, this statement might appear a contradiction to my entire theme of constructed ecologies, but I will explain my contention. There is a pressing need to distinguish fact from cultural opinions about the nature of Nature. This distinction is needed more clearly than ever because many design professionals reject science at the very time when the door is open for designers to join forces more strongly with engineering, ecological science, hydrology, physics, and mathematics. Yet landscape architecture is grappling with how to work with scientific knowledge and methods.

When I first came to landscape architecture from working in plant eco-physiology and mathematical ecology, I was presented with the alleged importance of pink bagels on a small lawn,[2] and to curious debates about whether nature was real. It seemed that only moments before my favourite book had been Martin Zimmermann's beautiful *Xylem Structure and the Ascent of Sap*[3] that describes the three-dimensional hydraulic architecture of woody plants. With this in mind, I also encountered a nagging problem in landscape architecture that arises from time to time. When I was a student, the landscape historian had set a small and delightful assignment of writing a 2,000-word play in which seven famous landscape architects were to discuss their design ideas over a meal. Nearly everyone in the class included André Le Nôtre, the great landscape architect of the French Baroque who had designed the garden of Versailles, dining at table. However, everyone else in the class portrayed Le Nôtre as vain, conceited, rude to those below him in status, and fawning to King Louis XIV, the 'Sun King'. I protested. No one had done the research to establish how Le Nôtre's contemporaries had described him – as humble, kind, who treated the boy who worked in the pot shed with the same courtesy as he did the king, a loving husband, a wise mentor, and the only person to address the king as Louis to his face.[4] Truth was a very different thing than supposed by many, but Truth did not appear to have concerned my fellow students; an arrogant Le Nôtre was apparently much more fun. And yet in design, we cannot ignore the truth of ecological processes and behaviours if we are to work in the world. To do so would be at our peril.

The environment does not respond to points of debate or preference. In the Japanese earthquake and tsunami of 2011, temples in Japan that long ago had wisely been placed above known tsunami ranges survived, but that knowledge had seemingly been forgotten by modern planners. The tsunami's range was not a point of debate between us and natural processes; towns built below that line were swept aside in moments. Natural processes do not discuss; the environment is not a human construct. While our policies, laws, and legislations change with our understanding of ourselves, nature's 'laws' do not change. For example, we can change policies, but we cannot ask plants to convert more sugars or produce more oxygen tomorrow, or ask a tropical plant to live in the Sahara. Nature is not a human construct dependent on our own limitless imagination, but is one of physical limits and finite riches.

It is important that design works with a recognition of the distinction whether we are dealing with knowledge with a cultural overlay, endlessly referred to as a cultural construct, or with real and specific information that is not flexible to mere opinion. While the latter is part of our culture because it is part of science, it is not as mutable as opinion and preference. I believe that a failure to recognise this distinction is a central impasse in design. This failure is holding up landscape architecture from moving out of its current limitations to engage more specifically and creatively with major environmental issues. Almost twenty years ago, James Corner wrote that the 'active life processes of which ecology speaks – are rarely paralleled in the modern landscape architect's limited capacity to transfigure and transmute'.[5] At the heart of Corner's assertion is the failure to distinguish between culturally loaded ecological information with ideas that are fluid and debatable, and information that is not culturally loaded and obeys physical, chemical, and biological laws and processes[6] that reveal to us information needed for wise design. It seems that this last ecology, or definite data, was forgotten in the last few decades amid an insistence that the world is a human construct.

As the great landscape historian Oliver Rackham noted, 'There is nowadays a tendency to regard the landscape as a mere artefact, the product of human endeavour, and to forget about Nature as the player on the other side of the board'.[7] We need to engage with this player by having stronger fundamental knowledge of the active life processes of which ecology speaks, for in doing so we can be transfiguring and transmuting *with* the player on the other side of the board.

Expressions addressing the sensitive clavichord's harmonies

There are a number of expressions that I have not used at all, or not much, because they are close to Diderot's sensitive clavichord in their anthropocentrism. These are 'ecosystem services', the 'Anthropocene', and 'climate change'. Designers (and ecologists) often cite ecosystem services largely unaware of the debates and concerns about this term in ecology, as in many concepts. My caution about ecosystem services lies in its general failure to deliver good science. In a paper entitled 'Have ecosystem services been oversold?'[8] Jonathan Silvertown neatly caught these concerns, including ethical issues about judging species. Who judges us? The commodification and then monetisation of ecosystem services have constrained thinking on our relationships with nature and are ultimately deeply anthropocentric. Fundamentally, nature is not an actor in the human construct of the money market based on commodities and cannot be accommodated in those markets as a member of the human team. Nature cannot bargain; while environmental scientists might bargain on behalf of nature, natural processes and systems cannot take part. The idea that nature has an intrinsic value independent of human use cannot sit within any concept of nature as a 'service' to us. The concept of ecosystem services has been used to connect with policy-makers, but it must be used with intellectual caution due to the parallel decline in discussing biodiversity and nature conservation with policy-makers, as the vague expression 'natural capital' has replaced these two terms. We must always remember that the only natural capital equivalent to a New Guinea rainforest is a New Guinea rainforest.

I have not used the term Anthropocene here because of the issues I will raise when discussing the savannah theory in Chapter 3. I defer using a term without going deeply into its meaning and complexity, and into the discipline in which the real decisions of that intellectual territory are made. In 2000, the Nobel atmospheric chemist Paul Crutzen suggested the 'Anthropocene Age' when casually talking to the press. Designers, everywhere, have taken it up, without pause. However, geologists – those who determine the ages of earth – are still discussing the term and have yet to pronounce upon it because they already have a term for the human era – the Holocene.[9] Many geologists feel that the data are insufficiently distinct and consistent (or dateable) to substantiate a Holocene/Anthropocene boundary (a 'global event horizon');[10] in short, the term might be used informally (with a little 'a') but not formally,[11] as the 'Age of', much like the 'Age of Steam'.

Eric Wolff, Royal Society Research Professor of Climate Change and Earth-Ocean-Atmosphere Systems, points out that the evidence is still coming in and it is best to let future generations decide on any potential naming of any Holocene boundary.[12] In particular, as is clear from what I have written, and from the references I cite, humans have been creating major disturbance on the world for 10,000 years. While more of the same is not a boundary, we might be witnessing an ethical and spiritual boundary, a boundary of 'giving up' – as many conservationists are profoundly concerned about – of any feeling of the Earth as our mother, as Lucretius so eloquently observed long ago. This is a dangerous 'theology' because although we alter, add to, and distort, we do not create the fundamental physical, chemical, and biological processes of the planet. Wisdom, including that sustained in indigenous societies, should remind us of this fact.

Nor have I explicitly focused on climate change in the modern era. Climate change has become the overriding narrative of the times, and that is dangerous. First, it blinds us to other important issues that impact ecology, human imagination, and knowledge, and second it is a non-discrete problem, has no elegant simple solution, and cannot be 'solved'. Population increase, poverty, global political tensions, global jihad (whose members care not for climate change, scientific inquiry, or humanist ideas), and pandemics rank as equally likely to unbalance human existence as a species. One failure of the politics and language of climate change has been a failure to see change as the normal condition of earth, or to discuss uncertainty.[13] Landscape designers need to manage change and flux in order to liberate our design ideas. Climate change is a constant companion, but I take the view of climate scientist Mike Hulme – in his superb *Why We Disagree about Climate Change* – that climate change is an opportunity for self-correction and rethinking, not the overwhelming disaster of the future. We need to unlock the uncertain future. We will make mistakes, but we always have self-corrected, and we will self-correct, as we are now doing in many fields. This attitude gives us all a far more positive future towards which we can work.

Problems of how we use ecology in design

Design has for too long taken two opposing views on how to consider ecology, views seemingly polarised between rigid empiricism and a loose choice, where one opinion is equal to any other opinion, whether informed or not. This problem has not done the design professions much good, leaving them locked in chatting, and not understanding phenomena or processes. The contrast between these two positions in the discipline of landscape architecture has been a microcosm of C.P. Snow's original 1949 long essay on the tyrannies of the two cultures – the arts and the sciences;[14] it has tended to place everything in apposition, rather than as a discipline employing the best available knowledge.

I have reflected upon ideas in this book that are either core to current design discussions, or missing from them. My approach focuses on knowledge gained from science that designers can use. The strength of this approach is that scientific knowledge is continually fluxing and growing, as greater understanding comes piece by piece – sometimes surging forward, sometimes edging backwards – providing new insights into how the world works. The scientific questioning that comes down to us from the Western Enlightenment has produced a globally unique culture of intense self-criticism and self-correction that still struggles against prejudice, rote-learning, zealots, lazy thinking, poor logic, and immobility of thought. Scientific questioning is essential to the understanding we now have about ourselves historically and organically as a species, as to how the physical and biological earth has changed repeatedly and continually, and how designs might perform well or fail.

Knowledge gives a strong structure to our actions and self-critique. To do this, people have to be well informed, with accurate data, information, and conceptual understanding. It is this knowledge that will assist us to not 'combat' climate change – because it is not a war with Mother Nature – but to amend, to shift, to accommodate change, and to create multiple strategies for any climate futures. In this book, there are no tales of woe of the future, nor any tales of doom and destruction due to the continuing Industrial Revolution for which all countries now clamour because such tales are not helpful to the great intellectual challenges in this and the coming centuries. Instead, I have set out to inform and reflect on ecology with design as self-critique as part of a longer tradition.

I set out a range of ideas to express ecological knowledge that can shift thinking in design away from 'ecological' opinion without facts, based on little information or knowledge. Sometimes designers argue an ecological opinion by citing major philosophers, yet this is perplexing because these authors are being used outside their expertise, are used without context, or add little of value.[15]

In contrast, explicit knowledge and rigorous inquiry from science give designers theoretical and positive concepts, new opportunities for specific design experimentation, and fresh ideas to imaginatively make and create. Because of this capacity, ecology – itself the study of intertwined systems – should not be seen as a separate 'bit' of design. Designers can gain major ideas to inform their designs from the concepts, graphs, tests, insights, and queries of the scientists who work in ecological science, conservation, mathematical biology, medicine, forestry, and agriculture.

What then is a constructed ecology?

Defining what a *constructed ecology* might be is a difficult thing, but I must make an attempt at the outset. The past-tense verb form suggests the act of construction and means that we took part in it, that constructed ecologies are those created in some manner by ourselves, and this definition lies behind this book. Why 'ecologies' – plural? We have always constructed ecologies; our ancestors constructed an ecology to create better hunting grounds or fishing sites. There are long-standing examples in the environment of how we made our own ecologies for our own purposes long ago, and today we are more extensively and self-consciously engaged in doing so, with equally determined endpoints, or ecological ambitions. In addition, ecology is complex; we measure and we miss other worlds within worlds, and I am acutely aware of this.

Constructed ecologies can be considered as falling into three broad categories or manners of creation: first, as sites of environmental history; second, as specific particular interventions carried out by engineers or designers as designed and constructed projects; and third, those ecologies inadvertently produced while we were busy doing something else, such as the side-impacts of dam building, farm management strategies, or suburban backyard plantings or paving. If we consider the images below, we can see that these artificial categories are blurred as they change across time.

The Fens (Figure 1.1), a large area of eastern England formed after the last Ice Age, are about 10,000 years old. They were naturally areas of low land, inundation, water, and wildfowl. In the seventeenth century engineers from Holland largely reconstructed these peat fens[17] as a response to the dangers of extensive flooding from the North Sea.[18] This region became one of the richest agricultural areas of England, producing over half of the Grade 1 agricultural land today. Recently, there has been a change as to how these highly constructed water-land systems will operate in the future. The reasons for this change go to the core of constructed ecologies. The revision arises from concerns about the loss of fenland and its rich capacity for biodiversity, the long-drawn-out lowering of the peat surface, the maintenance of riverbanks, and the capacity of the constructed fens to

Figure 1.1 The Fens in The Wash region of eastern England, near the River Ouse. The top image shows the drained fenland near King's Lynn, and the bottom, Wicken Fen, an example of surviving fenland. The mounds or hills in the top image show the previous level of the land, that is now a rich cereal and vegetable farming region (e.g. potatoes, carrots, sugar-beets), all growing on fertile peat soil. The original fen in the lower figure was, in the twelfth century, abundant in wildlife, with 'such a quantity of fish as to cause astonishment in strangers',[16] and provided wood, thatch, fodder, birds, and fish for human use.

Images: The author.

act as a buffer to North Sea storm surges, combined with rising sea levels due to climate change. Importantly, we are revising our very relationship with how we deal with the edge – the relationship of water and land. Out of these concerns in the 1990s has come the Great Fens reconstruction project to restore and reimagine the Fen country of England. The story of these fens shows a startling series of constructed changes to one landscape; it is environmental history, a story of deliberate change, and one of unanticipated side-effects.

A synopsis of the book

Constructed Ecologies consists of six substantive chapters, each an essay on core areas of landscape thinking. In Chapter 2, 'Global differences, not universals', I examine global generalisations. A tension in ecology is the search for globally applicable hypotheses, as if the global is always better than the local. Such tension is key to design because designers need to design at the local scale, but with their eye to 'big-picture' ideas. Simplification into generalisations rarely accommodates the diversity of our world, whether cultural or physical. I introduce the ideas of *spectrums of responses* and *shifting continuities*.

In Chapter 3, 'Shifting adaptabilities, not static concepts', I reappraise what the landscapes of early modern humans were from research in palaeobotany, anthropology, neurology, and cognitive archaeology. In this chapter, I tease apart the savannah theory, a well-known paradigm in landscape perception research that has become a default statement of how humans perceive landscapes, leaving both researchers and designers with a limited criticality. As I reappraise our early landscapes, I show that we lived in multiple environments and that, as climate changed over the last 70,000 years, we often stayed where we were and adapted to changing landscapes while environments around us moved from forest to savannah back to woodland. The one constant in human history was change. I suggest a new approach – of *shifting adaptabilities*. This shift in concept has positive implications in our ability to adapt to any climate changes and is directly related to our current ecological imperatives and human capacities for artistic and imaginative responses. I suggest that this proposed new paradigm will likely lead to a revision or re-examination of the current ideas of landscape perception and lead to a reawakening of design ideas for public spaces.

In the second part of the book, I scrutinise some important ideas that lie behind design thinking. Chapter 4, 'Multiple, not solo voices', continues the theme of change and flux, with an examination of multiple imperatives, with discussion focused on agriculture and design. Many ecologists, engineers, and designers will be unaware of substantive changes in agriculture, one of the great cultural impetuses of the world and the greatest constructed ecology of the planet. Many in the developed world and developing world are reimagining and reworking agricultural production and the farm, with many additional imperatives to food production. Can designers assist in the major human challenges associated with farming and move beyond ideas of urban food production and local food initiatives in cities? In this chapter I suggest a new and very specific landscape architecture arena of *design 'georgics'* – the 'things of the farmer'.[19]

In Chapter 5, 'Inquiries, not assumptions', I discuss a little-considered constructed ecology, artificial light at night, that impacts ecological processes, human health, and carbon, and is a major energy cost of the urban landscape. Despite these wide impacts, designers neglect this area of inquiry, with design and planning our night landscapes led by closely held assumptions and dominated by the lens of energy saving. New information suggests that artificial light at night is implicated in increased cancers, sleep disruption, and the death of animals. How can the design community respond to new information, and redesign?

In Chapter 6, 'Thinking backwards, not forwards as a linear narrative', I show that there is no one method for an outcome, no linear route, and that there are multiple possibilities of what can be done to construct an ecologically sound outcome. This is moving design away from the ideas of Ian McHarg's *Design with Nature* and the overlay technique that has influenced design for decades. As I reframe design as an inverse problem of questions and multiple answers, with strong ecological, cultural, and artistic outcomes, I hope that asking the questions about how we are designing will assist inventiveness in design.

Within this book I propose inquiry and interrogation. Some ideas might instil a new sense of purpose, while others might suggest new routes to members of the design professions who might find their niche for future work.[20]

Notes

1 Denis Diderot, 1769, 'D'Alembert's Dream', p. 105, in *Rameau's Nephew; and, D'Alembert's Dream*, translated with introductions by Leonard Tancock. Diderot was one of the Encyclopaedists.
2 As noted by Richard Weller, 2014, 'Wordscape: The writings of James Corner in theory and practice', Afterword, in James Corner and Alison Bick Hirsch (eds), *The Landscape Imagination: Collected Essays of James Corner 1990–2010*, New York: Princeton Architectural Press, pp. 351–361.
3 Martin H. Zimmermann, 1983, *Xylem Structure and the Ascent of Sap*, Berlin: Springer. Xylem are the major structures that transport water in plants, and they have held a long fascination among botanists as to how plants take up water to the highest branches.
4 See, for example, James Eugene Farmer, 1906, *Versailles and the Court under Louis XIV*, New York: The Century Company, pp. 81–83.

5 James Corner, 1997, 'Ecology and landscape as agents of creativity', in Corner and Hirsch, 2014, p. 261.
6 See, for example, Mark Denny, 2016, *Ecological Mechanics: Principles of Life's Physical Interactions*, New York: Princeton University Press. In this book Denny unites ecology, physics, and engineering, an approach that points the way to a similar union in landscape architecture for students with personal proclivities toward these arenas. Denny makes the point that it is surprising how far a little bit of mathematics or physics learnt at school can take a learner.
7 Oliver Rackham, 1979, 'Documentary evidence for the historical ecologist', *Landscape History* 1(1): 29–33.
8 Jonathan Silvertown, 2015, 'Have ecosystem services been oversold?', *Trends in Ecology and Evolution* 30(11): 641–648; and Silvertown, 2016, 'Ecologists need to be cautious about economic metaphors', *Trends in Ecology and Evolution* 31(5): 336. See also useful comments on Silvertown's papers in Kerrie A. Wilson and Elizabeth Law, 2016, 'How to avoid underselling biodiversity with ecosystem services: A response to Silvertown', *Trends in Ecology and Evolution* 31(5): 332–333.
9 P.L. Gibbard and M.J.C. Walker, 2014, 'The term "Anthropocene" in the context of formal geological classification', in C.N. Waters, J.A. Zalasiewicz, Mark Williams, Michael Ellis, and Andrea M. Snelling (eds), *A Stratigraphical Basis for the Anthropocene*, Geological Society Special Publication 395, London: British Geological Survey, pp. 29–37.
10 Mike Walker, Phil Gibbard, and John Lowe, 2015, 'Comment on "When did the Anthropocene begin? A mid-twentieth century boundary is stratigraphically optimal" by Jan Zalasiewicz *et al.*', *Quaternary International* 383: 204–207.
11 Gibbard and Walker, 2014.
12 Eric W. Wolff, 2014, 'Ice sheets and the Anthropocene', in Waters, *et al.*, *A Stratigraphical Basis for the Anthropocene*, pp. 255–263.
13 For an excellent discussion about everything in relation to climate change, see Mike Hulme, 2009, *Why We Disagree about Climate Change: Understanding Controversy, Inaction and Opportunity*, Cambridge: Cambridge University Press. While the main title is ambiguous, and perhaps misleading as to the contents, the book sets out the approaches to how people think about climate change. Hulme asks not what we can or must do about climate change, but what climate change can do for us, with ambiguity and uncertainty as resources.
14 C.P. Snow, 1998, *The Two Cultures*, Cambridge: Cambridge University Press. This book caused a great deal of interest when first published in 1949 and continues to do so.
15 As an example, Deleuze's *Three Ecologies* is not about ecology, but I have seen this work cited as if an important text on the subject; it is not.
16 William of Malmesbury, *De gestis pontificum Anglorum*, written about 1125, in H.C. Danby, 1940, *The Medieval Fenland*, Cambridge: Cambridge University Press, p. 28.
17 See Michael Chisholm, 2012, 'Water management in the fens before the introduction of pumps', *Landscape History* 33(1): 45–68, for a discussion of the development of the fens and the major re-routing of rivers in this region of England. Henry C. Danby (1956) gives a full description of the history of the fens in *The Draining of the Fens*, Cambridge: Cambridge University Press. The newly drained fens can be compared with the old fenland in his companion book *Medieval Fenland*.
18 See also Margaret Knittle, 2007, 'The design for the initial drainage of the Great Level of the Fens: An historical whodunit in three parts', *Agricultural*

  History Review 55: 23–50; and Michael Chisholm, 2012, 'Water management in the Fens before the introduction of pumps', *Landscape History* 33(1): 45–68.
19 From the Greek translation.
20 In Margaret J. Grose, 2014, 'Gaps and futures in working between ecology and design for constructed ecologies', *Landscape and Urban Planning* 132: 69–78, I discuss the differences between design and ecology, the relationships between these disciplines, and new routes of collaboration.

2

GLOBAL DIFFERENCES, NOT UNIVERSALS

Reason respects the differences,
and imagination the similitudes of things
        Percy Bysshe Shelley, *A Defence of Poetry*, 1821

Here we have two images – one of the Sphinx and the other of common beech (*Fagus sylvatica*) in England, where it is cherished as a beautiful native tree.[1] The Sphinx was built in about 2,500 BC, approximately 4,500 years ago. I ask my students which has been in the ground in place longer – the Sphinx in the sands on the edge of the Nile Valley, or the species we call beech in England? The answer is the Sphinx. Common beech arrived in the United Kingdom only about 3,500 years ago,[2] which is 1,000 years after the building of the Sphinx and only 250 years or so before the Trojan War. If we are surprised by this, it is likely because we are used to thinking of native trees as longstanding components of a continual ecosystem, or because the answer is not what we might expect of biology compared with built form. We might think this because the fabulous development and vast speciation of plants has led to a presumption that they are aged and fairly fixed. Yet plants and animals have moved in space. This chapter is essentially about plants and their past and future movements and about ideas for design that arise from knowledge of plant movement.

Since the articulation of plant structure, taxonomy, and function in the eighteenth century, we have assumed that plant species belong in some certain geographic location. As the great geographer and botanist Alexander von Humboldt wrote in his descriptive book *Views of the Cordilleras and Monuments of the Indigenous Peoples of the Americas* in 1810–1813: 'As one changes latitude and climate, one notices a change in the appearance of organic nature, the shape of the animals and the plants, that lends a particular character to each zone.'[3] This was a beautiful idea, and botanists mapped the world's climate zones, biomes, and plant distributions and types, such as tropical forest or temperate woodland, as patterns on the landscape. When Humboldt drew the delightful representation of the Andes in 1805 (Figure 2.1), we knew where plants were and learnt how they grew in those defined places and what they needed climatically. There is a known tree-line where the average summer temperature does not exceed 10°C; there is a tundra line where its treelessness separates it from the conifers of the taiga; there is a fuzzy region where deciduous broadleaf trees give way to evergreen needle forests; there are Köppen climate classifications, and we know what climate type we grew up in; mine was Mediterranean.

All of this mapping was based on the known relationships between place, climate, and vegetation. Learnt at school, this readily available geographic knowledge gave us certainty, and we learnt to use that certainty to consider the world and to formulate our ideas about constructing plant ecologies in that world. We presumed that the suites of plants that grew together would remain together in their ecosystems and plant assemblages. Now, with current climate change predictions for the coming centuries and the rapid changes in this century, we must throw all of that knowledge into the air and reconsider plant distribution. However, the great driver of this changed view of the world has not come from our views of the future, but from our ever-increasing understanding of the past. As palaeobotany, palaeo-phytogeography, and macro-ecological history reveal previous geographic distributions of plants, they are suggesting to us that future plant and animal distributions will be different to current maps and zones and that the future is largely uncertain. New knowledge is throwing aside our old certainty about plants and place and is breaking that long-nourished link between climate, place, and vegetation. The major question for us as designers is: how will we address this uncertainty, and what issues will uncertainty bring us? Can our constructed landscapes assist plants and animals to survive and prosper?

Figure 2.1 The image shows plants found, at certain locations, on a gradient up a mountain in the Andes from the expedition of Alexander von Humboldt and the botanist Aimé Bonpland of 1799–1804 into Central and South America, *Geographie des Plantes Equinoxiales. Tableaux physique des Andes et pays voisin.*

Source: AKG images, with permission.

In this chapter I discuss our thinking about place and species in design, now and in the future, and propose the two ideas – *spectrums of responses* and *shifting continuities*. By *spectrums of responses*, I mean the range of possible ways in which we might design with plants that is dependent on their relative evolutionary histories, their rarity or commonness, the support they give to other species, the genetic diversity of individual species, and their capacity to adapt to new conditions. In the expression *shifting continuities*, I refer to the need to maintain plant populations and species in the face of changing climate and other shifting conditions, including urbanisation, which will put spatial pressure upon plants. This term refers to design, planning, and conservation actions to enable their continued survival. The concepts of *spectrums of responses* and *shifting continuities* are needed in order to consider a whole host of ideas about plants and their landscapes – native and non-native, novel ecosystems or third nature or no-analogue communities, conservation and restoration, mass planting of single species or massed genetic diversity, refugia and micro-refugia, design for future continuities with climate change, 'rewilding' of landscapes, and children's natural play and rewilding playgrounds; these areas of study and design are all interrelated.

I ask here how macro-ecological history – the history of large-scale plant and animal changes – can inform design thinking and practice. I delineate the macro-evolutionary history of the northern landmasses and the southern continents and their associated phyto-geographic[4] histories, and discuss potential design implications of their different histories. That the continents have different legacies from their histories is important because both science and design remain caught in a 'one-size-fits-all' global approach, with a search for commonalities or a presumption of commonalities. This is the very opposite of ecological thinking, which stresses the importance of site specificity in landscape responses, processes, variation, and scale that will produce varying responses of place and time, not a one-size-fits-all. The large-scale differences in the macro-ecological histories of the continents are the starting points for how we might consider plants and design.

Hemispheric differences:
Northern movement and loss, and Southern survival

In the last 1.9 million years, there have been several glacial periods during which ice sheets covered much of Earth's high latitudes and several shorter interglacial periods of warmer conditions. The most recent driver of changes between the continents, and the last major climatic disturbance of plant distribution on Earth, was a glacial period that began about 25,000 years ago, came to a maximum extent – known as the 'glacial maximum' – about 18,000 years ago, and finally dissipated about 11,700 years ago. The manner in which plants responded to this enormous climatic perturbation of increased cold and reduced precipitation – creating ice, low sea levels, and regions of great aridity – has been the subject of intense inquiry in the discipline of palaeobotany. This information assists thinking about design for climate change.

The last Ice Age left different legacies in the Northern Hemisphere and the Southern Hemisphere and within each hemisphere. Physical barriers such as mountains, and the new deserts of the Ice Age created profound disparities in the levels of extinctions and plant survival. Europe lost a high percentage of its flora, while North America had relatively few extinctions.[5] The stark contrasts in these extinctions can be attributed to the latitudes of the continents, their extent or continentality, and the alignment of mountain ranges. In Europe the mountain ranges – the Pyrenees, the Alps, the Carpathians, the Balkan Mountains, and the Caucasus – are aligned in a mainly west–east direction.[6] In Asia the mountains of the Hindu Kush, the Greater Himalayas, the Tien Shan, the Altai, and the Kunlunshan provided an impenetrable barrier to movement south. With ice expanding southwards during the Ice Age, these east–west mountain ranges 'trapped' plants migrating south in Europe and Asia as areas north of the mountains became extensive plains of permafrost, tundra, and cold steppe.[7] As plant distributions moved south in response to increasing cold and reduced growing seasons, the land was increasing in altitude and getting colder, not warmer. With nowhere to go, extinction followed for many plants and animals. In Europe, the Mediterranean Sea provided another insurmountable barrier to plant retreat from increasingly cold European conditions.

However, in North America plants had no obstructions to moving south with increasingly cold climatic conditions because the mountains in North America – the Rockies, the Appalachians, and the Sierra Madres – align north–south. At the height of the last glaciation, spruce (*Picea* species),[8] today found only at high latitudes, was a common tree in what is now Louisiana,[9] in extensive plains that were out to sea due to the lower sea levels at that time. Of course, it is not correct to think of plants as 'migrating' because they cannot pack their bags and leave, and the creeping arrival of ice meant massive death *in situ* for millions and millions of individuals. It was their seed that migrated bit by bit across the landscape. If the seed failed, the species failed.

In contrast to the severe glaciation in the Northern Hemisphere, glaciation in the Southern Hemisphere was short and had minimal impact. In Australia there was little glaciation due to its low elevation and temperate latitude, and Australia maintained its earlier legacies on isolated landscapes with a strong and unique biological diversity,[10] but experienced increased aridity. New Zealand experienced glaciation, and in Africa extreme aridity was the major feature. In South America glaciation occurred along the Andes but, with north–south running mountains, plants were able to retreat, migrate and survive, as in North America.

Clearly, plant responses to the last Ice Age were not universal. Yet despite differences between these histories in the hemispheres, there is a strong tendency to presume that studies or design projects done anywhere are relevant to everywhere. The Northern Hemisphere, with its young physiography, landscapes, and strong glacial impacts, dominates the landscape architecture literature. Designers need to be mindful that work done there might have no relevance to other regions. Global differences exist in palaeohistory, not universals, and designers can explore these characteristics as a source of design inspiration.

It is important to remember that humans journeyed far during the Ice Age also. In Australia, living stories remain of our experience. For several years I have taken students into the 'High Country' of Australia – to the Snowy Mountains and the Kosciuszko National Park. This is the world's most climatically sensitive alpine area because it is comparatively low, having been softly eroded over millennia. Australia's highest mountain, Mount Kosciuszko, is 2,246 metres (7,368 feet), and the area is predicted to be ice-free by 2050. With a warming climate, plants and animals that require cold in this region will have 'nowhere uphill to go' – they cannot go higher because higher does not exist.

Mount Kosciuszko is the centre of a spirit world for Aboriginal peoples. Songlines lead from it to many places in Australia, even to sites very distant. In 2014, my students were privileged to meet the great soul, Mr Rod Mason, an Elder of the Ngarigo tribe, whose Country is the Snowy Mountains region. Rod related the story that when the ice came his tribe's life in the high country became untenable, due to year-long snow and biting cold. They made their way inland to sites near Kata Tjuta in central Australia. Every generation, a party of men travelled to see if the ice were still there; it remained for a long while. Finally, after many generations, the scouting parties reported that the ice was beginning to melt away in the season of warmth; the tribe kept close watch and returned to their country when wildlife began returning. We are privileged still, to have this extraordinarily precious record of human life from before, during, and after the last Ice Age.

Retreat and Ice Age refuge
is a particular legacy with us today

During the last Ice Age, plants that had successfully arrived in sites away from the ice where they could reproduce successfully remained in those refuges. Refugia is a term referring to areas that may facilitate the persistence of species during large-scale, persistent climate change.[11,12] Northern Europe was covered with an extensive and continuous ice sheet while southern Europe was cold,[13] but acted as a refugia for the current European flora. Most of the organisms present today in Europe survived in the peninsulas of Iberia, Italy, and the Balkans, while some moved far to the east on the edges of the Caspian Sea.[14]

Refugia are not confined to glacial periods. Cold-adapted species that were more extensive during the last Ice Age have been constricted during the current, warmer, interglacial to colder areas that act for them as refugia today in our warmer conditions. Examples in Europe are found in northern Scandinavia and the Alps.[15] These cold refugial species will be under enormous pressure in the coming decades. In this way, the concept of refugia is not one locked into palaeohistory, but is of use today in current climate change. The distribution of some plants will spread with a warmer climate; others will retreat; some will have nowhere to go.

Within refugia, localised buffering from extreme environmental effects led not only to protection of many species from extinction, but also allowed hybridisation of species. That is, it allowed for the development of new species and the development of hotspots of endemism for both plants and animals.[16] We can ask – where might refugia be possible in the coming centuries?

Dispersal after the Ice Age, and the questions it raises for designers today

About 16,000 years ago, as the climate warmed, ice began to retreat to its approximate present position. During this retreat, plants dispersed into areas that were previously too cold or ice-bound. This was not an evenly spread event. Species moved at different rates, with different mixtures of plants meeting and then dispersing – as one species was able to move on, while another was delayed, or stopped. Communities of plants were, therefore, not stable, and there was neither a permanent species assemblage,[17] nor species assemblages that moved in concert. The palaeobotanist Gordon Hewitt notes, 'Species responded individually to this warming with each tracking their particular set of environmental conditions. This created mixtures of species different from today, that were transitory so that communities were not stable.'[18]

The journeys that plants took into their present positions are being tracked by palaeobotanists. Each track reveals a different response to warmth, soils, opportunities for seed distribution, and the vectors required for seed dispersal (birds, gravity, digestion, wind) for that expansion. Particularly important were the various responses to the defined growth margins of species for seasonal temperature.[19] These transitory communities of plants were novel ecosystems, and this historical shifting is a reason why novel ecosystems are not new phenomena on Earth.[20] Plants travelled at both the population level and as groups or individuals. Using techniques of pollen analysis and macro-fossil data (plant remains apart from pollen) and examining chloroplast DNA, we can see the impacts of various refugia, as well as the travels and shifting place-associations of plants. Examples of these massive dispersals (Figure 2.2) show the changing ranges of trees in North America over time.

Trees are particularly interesting to follow in any story about change and movement with climate pressure. Their generally long-lived nature suggests that their generation-time, or the change of genes, will be slow – an idea first put forward a century ago[21] – and will therefore be vulnerable to rapid climate changes. Trees usually have high rates of genetic diversity, and genetics can reveal where trees have come from and where they have travelled since the last Ice Age. Their high levels of genetic diversity mean that some regions that support the same species have genetically different trees – still outwardly the same species, but genetically marked by their history of coming from refugial populations that were once genetically isolated during the Ice Age.[22]

The European beech (*Fagus sylvatica*) at the front of this chapter is a good case study of what happened to trees during the return of species to land opened up after the retreat of the ice. Botanists have examined the macro-ecological history of beech because it is one of the most important economic trees of Europe.[23] There is a good pollen record of beech's post-glacial advance,[24] which was relatively slow compared with other more widespread trees like oak. While oak reached Britain about 9,500 years ago,[25] *Fagus* reached England between 3,500 and 2,000 years ago after expanding in fits and starts at various speeds across the European landmass. Botanists refer to this as a heterogeneous population expansion, similar to the movement of many tree species in North America after the Ice Age.[26,27]

The complexity of plant movement and the subsequent differences in genetic legacies today appear in work done in Europe on the European white oak (*Quercus* species). White oak is readily studied because it is common from northern Africa to the Caucasus, and is a key species in ecosystems. Oak trees produce a lot of pollen and therefore can be readily tracked as to where they have come from historically. Because oaks have low levels of in-population diversity at any particular site or forest, only a few trees need to be sampled to get a picture of

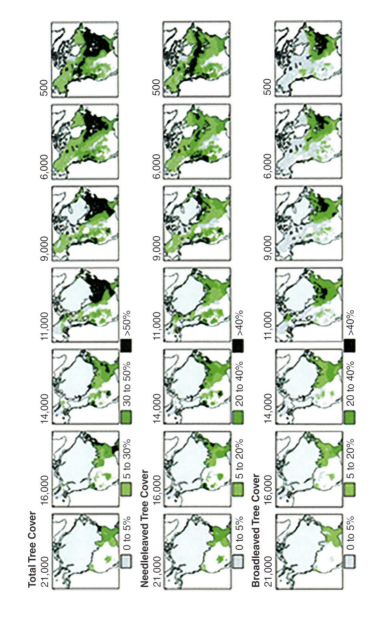

Figure 2.2 Here, in a sequence from 21,000 years ago to the near-present, we can see how plant species have been in motion. The example shows maps of reconstructed tree cover over the last 21,000 years in North America; progressively darker green shading indicates increasing abundance of that type (i.e. greater density): total tree cover, needle-leaved trees, and broadleaved trees. Grey indicates absence and white no data. Details of this figure are adapted from: John W. Williams, 2003, 'Variations in tree cover in North America since the last glacial maximum', *Global and Planetary Change* 35: 1–23.

Image: Permission from Elsevier.

their genetics. Indeed, patches of several hundred square kilometres exist that 'are virtually fixed for a single haplotype'[28] – in other words, with very little genetic variation. Studies across Europe amongst sixteen labs and more than 2,500 populations of white oaks have established where the main haplotypes exist today, and where they had taken refuge during the Ice Age.

Examples of this research were carried out using chloroplast DNA analysis (Figures 2.3 and 2.4).[29] Chloroplast DNA tells us a great deal about where different groups (populations) of the same species emerged geographically after the retreat of the ice. Such data also indicate at what speeds that population then moved once warmer conditions arose. In Europe, these last great travels for plant species occurred at rates ranging between 150 and 500 metres per year for most trees, with some achieving more rapid recolonisation, and others slower.[30]

Chloroplast DNA reveals to us that while an individual tree of one species might look the same as any other, it might be genetically different. A single tree's genetics will depend on the Ice Age refuge where the population lived and survived. Work on plant DNA raises the question of whether preserving genetic diversity in trees is important – and we should presume that it is for future biological security – and from where do nurseries get their stock? Is stock regionally specific or do designers and horticulturalists take the attitude that anything available will do, because availability is important in completing a designed project on time? We might be tempted to take plants from a nursery that offers a better financial deal if a local supplier does not have stock, or if it seems too expensive. However, if a designer wants to plant 120 white oaks (*Quercus robur*) in Scotland it might be important to obtain a haploid type that is found in Scotland, and not one found primarily in Estonia but never found in Scotland. Is it an error if we take the non-Scottish one? Is there an ethical aspect to acknowledging haploid variability in planting? Many ecologists would argue that there is, because haploid variability is a part of biodiversity, in this case molecular genetic diversity, that maintains the rigour and genetic potential of a species. Concerns about biodiversity at the genetic level also reveal the value of regional specialists in plant material, with regionally specific genetic plants being important. The argument against keeping genetically different species separate in geographical space is that a designed population with a wide genetic background will be more resilient, so that it is more likely that some members of the population will survive a serous perturbation.[31] This debate is ongoing.

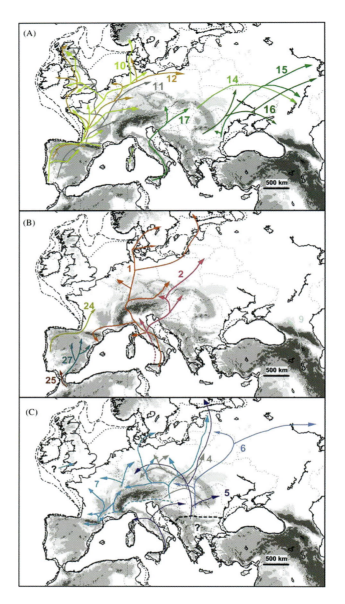

Figure 2.3 This figure shows the movement of white oak, based on evidence from chloroplast DNA and fossil pollen, across Europe following the end of the last Ice Age, for various haplotypes given by numbers in all three maps. Petit *et al.*, 2002, also indicates the altitude: 'by grey shadings (250–500, 500–1000, >1000 m) and past sea levels at 21 ka BP, 15 ka BP and 12 ka BP are indicated by dotted lines'. For details see Petit *et al.*, 2002.

Image: Reproduced with permission from Elsevier.

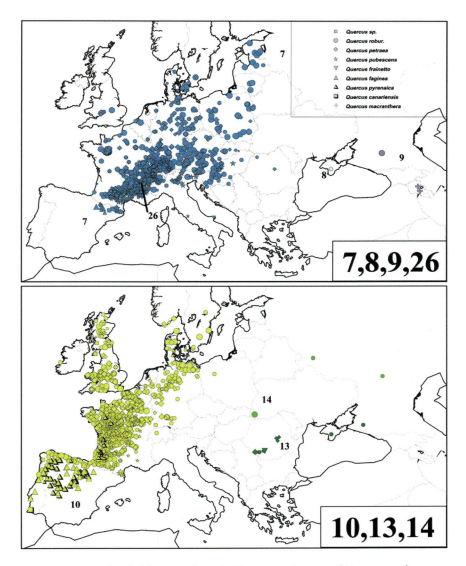

Figure 2.4 Here, haploid groups 7, 8, 9, 26 are seen in central Europe and eastern France and into Poland, in comparison with haploid groups 10, 13 and 14. Haploid group 10 is found only in western France, Britain, and Spain. This raises the issue for designers of where plant stock comes from, and what haploid group to plant. For details of this and various other haploid groups, see Petit *et al.*, 2002.

Image: Reproduced with permission from Elsevier.

Spaces and places

Mapping the travels of the world's great tree species shows the genetic diversity of their various populations as they are distributed today. Historic dispersals are an indication of where plants might be moving in future warmer conditions. The greater understanding of plant change in response to Earth's climate history turns on its head that old link between species and place. Geographic mapping reflected but a moment in time, when we just happened to be recording the data during a few hundred years of detailed science. But we captured only one point in a dynamic process of changing relationships of plants to place; these were temporary links, shifted in history, still shifting, and about to shift again as we take note and record them. The rate of change will not be constant over all hemispheres, continents, regions, places, valleys, plateaux, or smaller topographic sites. The dynamic rate of change might be fast in some places, slow in others. For example, spruce is invading the tundra now while, in other parts of the world, some species remain where they have been for tens of thousands of years, and some are likely to remain in place in the coming centuries.

Landscape designers have rarely drawn ideas from the genetic history of plant species. We still rely on the snapshot of plants and place and the stable geographies of plant distribution,[32] that in turn has focused on what we might plant, and where. In contrast, the genetic history and the lessons learned from our knowledge of the last Ice Age point us to where design inquiries might take us.

Sharpening landscape planting through the lens of historical plant dispersal[33]

The patterns of plant dispersal differed, and knowledge of this aspect of macro-ecological history might inform design choices in the hemispheres. The great macro-ecologist Godfrey Hewitt noted a particular legacy as being 'southern richness, northern purity'.[34] He considered that the northern parts of Europe have large land areas of near monocultures, such as occur with larch and spruce, two species that occupied sites very quickly after the retreat of ice and tundra; in southern Europe greater biodiversity is found within sites. The same relationship can also occur in North America and Asia, with northern purity in the northern latitudes, and greater diversity generally in the south, where mosaic habitats and stable topography also played a part. We might also consider 'northern purity and southern richness' between the hemispheres, where great biodiversity occurs in the southern latitudes, such as in the Cape Floristic Region of South Africa and the Southwest Australian Floristic Region,[35] compared to the Northern Hemisphere.

GLOBAL DIFFERENCES, NOT UNIVERSALS

Could landscape architects explore the concept of 'northern purity, southern richness' by using it as a tightening frame, or by juxtaposing massed planting and complex richness? Does it mean that we might consider mass planting of low diversity in the northern latitudes, including Germany, Russia, northern China and northern North America? Should we consider massed genetic diversity in southern latitudes such as Greece and Spain, southern Australia, southern South America, and southern Africa? Might 'northern purity, southern richness' suggest a way of working with macro-ecological ideas to create spatial form, whereby we might arrange or visualise our plantings in tune with our particular continent's macro-ecological legacy, thereby removing homogeneity in global approaches to planting? Can we juxtapose these two approaches? Might this idea be too confining a nuance, or one that sets parameters, discipline, rigour, and direction in what choices we make in regard to both plantings and spatial arrangements? At what scales might we design? Might it mean that horticulturalists in the southern areas would have more fun?

The designer Michel Desvigne created work in Minnesota that appears to directly reference the spatial patterns of massed species monocultures of that northern climate (Figure 2.5). In the Walker Art

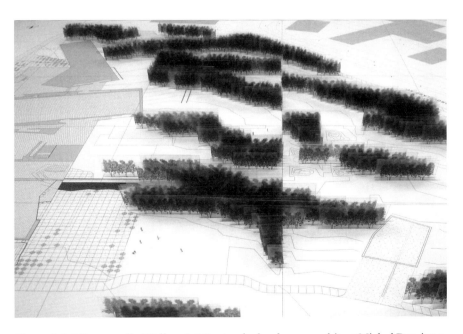

Figure 2.5 Minneapolis, Walker Art Center, by landscape architect Michel Desvigne.
Image: With permission MDP Michel Desvigne Paysagiste.

Center in Minneapolis, he arranged regular swathes of birch trees (*Betula nigra* and *Betula nigra* 'Heritage') and aspen (*Populus tremuloides*), with simple prairie grasses, to evoke the Midwest post-glacial landscape. This is the landscape equivalent of picking up the particular features of glaciated landscapes, the flow and sliding of glacial ice, and the detritus within the moraines left behind on a flattened land, of the much-celebrated prairie-style houses of Frank Lloyd Wright in the early twentieth century that reflected the palaeohistory of the prairies.

Projected futures of plant movement

How quickly did change occur in the last de-glaciation when the last Ice Age ended, and does the pace give us ideas about what will happen now? We cannot know precisely when and how quickly climate will change. Recent work in the Mediterranean has indicated that palaeoecological data can assist the projection of vegetation patterns or re-establishment following dispersal with climatic change.[36] However, there are no ready generalisations, no global universals, except that all species will be shifting according to their own needs, and many are likely not to shift at all. In some regions, previous climate change was very abrupt, with a period of free migration, with limits set by the time each new generation could produce seed and the distance that seed could be dispersed. Plant movement is hard to calculate because trees live a long time and lag times exist. Already in the Northern Hemisphere's boreal forest biome, warmer and shorter winters are disrupting current Arctic terrestrial ecosystems,[37] and scientists are recording northern shifts in optimal conditions for Arctic plants and animals.[38] Temperature is driving this transformation, and boreal forest regions in Canada and Russia are set to see major alterations in productivity, disturbance regimes, and carbon sequestration.[39] It is expected that changes will be heterogeneous.[40]

Scientists are projecting changes in tree compositions in the coming centuries. Jacqueline Mohan and her co-workers in the USA and Canada predict that tree species now in north-eastern USA will alter markedly.[41] They expect that currently common north-eastern species – balsam fir, paper birch, red spruce, bigtooth and quaking aspen, and black cherry – will decline, with a retreat of the spruce-fir zone into Canada.

Shifting of species is already being seen in higher northern latitudes. Although spruce lived in Florida during the last Ice Age, spruce now resides in the taiga, that vast biome between the tundra and the boreal forest to the south. The taiga is the largest biome in the world, and contains about 30 per cent of the world's forests across Russia and Canada. Spruce is now invading the northern tundra in the Yukon, Canada, as seen in Figure 2.6. Spruce is moving north with a speed that has staggered foresters and botanists; we might neither know nor

Figure 2.6 Pioneer spruce invading tundra, Burwash Uplands, Yukon, Canada. Transformations of landscape occurred in Europe, Asia, and North America following the warming at the end of the last Ice Age. Plant movement will increase in the next centuries. Here, a community of spruce trees appear to be in transit – but where is their destination this century?

Image: With permission from Ryan Danby.

appreciate the travel rates of even major tree species. Nor do we know what plant and animal species will be travelling with major tree species.[42] Ryan Danby, forest ecologist at Queen's University, Canada, notes that changes in tree and animal distribution will impact land use, the distribution of sheep and caribou, and will ultimately impact the lifestyles of Alaskan Natives and First Nation peoples.[43]

Designing for future climates: Enabling shifting continuities

In the face of climate uncertainty and species rearrangements, how will we make planting choices? What we are looking for is a way to ensure *shifting continuities* of species in the face of the general uncertainties of exactly how fast and where changes will occur, if at all. What we are trying to avoid is the death of future continuities,[44] such as occurred in the extinction of plant and animal populations with the last Ice Age, when many species did not find a refuge.

Can designers, planners, ecologists, and conservationists help continuity? It is likely that there will be a number of 'lines of defence' we might develop to allow plants to adapt to climate shifts. Some versatile lines of defence might be widely adopted, but we need a breadth of responses. There will be no prescriptive remedy for assistance; what might work in Florida might be useless in Maine; there will be global, regional, and sub-regional differences. Design can allow connectivity by corridors, increased habitat spaces, holdouts to enable plant adaptation if that is possible, small habitat intrusions, pushes for legislative changes to tree protection, water reuse in our vast suburbs and sprawling cities, and many other ideas not yet conceived. Such diverse strategies are needed because if we rely on one or two, and they fail, our ambitions as enablers of plant survival for future landscapes will be dashed.

How quickly must designers act to allow for climate changes? Will models of species movement and the most likely distributions and abundance of various species in the face of climate change impel designers to rework protected area networks, plant for the future climate not the present one, and work with conservationists more pointedly? The key point is that change in distribution, long associated with geography, is not a spatial problem but an ecological one.

Will national parks need to be revised and relocated, Green Belts replanted, and hotspots redefined? About 10–15 per cent of the world's land surface is under some form of protection, and these are all geographically fixed,[45] and based on the knowledge of plant and animal distribution of the last few hundred years. Important changes will occur in national parks. With climate change, national parks might not have the right landscape conditions for the preservation of what we are hoping to preserve there.[46] Many national parks are located on sites of natural beauty and rich flora and fauna, but are often surrounded by landscapes that are not hospitable to the species living within them, and species will again be 'trapped'. Likewise, the current designation of global hotspots of biodiversity, such as the California Floristic Province, home of both the Sierran giant sequoia and the coast redwood, might become 'Coolingspots', sites of moderated species diversity, or sites of species loss. Hotspots might not be adequately placed to allow for climate change.[47] Landscape architect Richard Weller and his associates at the University of Pennsylvania are conducting a mapping exercise to address this issue, scoping the ecological networks required to maintain protected areas.[48]

Does thinking of the macro scale allow us to consider how plants entered a new area after the last Ice Age, how they were replaced, how they overtook sites, and how some remained behind? Does thinking about shifting and fluxing allow us to consider time, accumulation, the density of species, and plant mixes? We now realise that plants and animals were just 'passing through'[49] and are shifting again. The idea of shift and flux is far removed from earlier ideas of places linked to plant species seen in Humboldt's lovely drawing. Indeed, the ideas of shift and flux have been around for some time in the biological community. How can we design for shift and dispersal of plants?

I must comment about the concept of novel ecosystems because it has been an immediately attractive one among designers. The expression novel ecosystems refers to a poorly defined and shifting idea that has arisen in ecology and refers mainly to 'a new species combination that arises spontaneously and irreversibly in response to anthropogenic land-use changes, species introductions, and climate change, without correspondence to any historic ecosystem'.[50] Many ecologists have criticised this term because it merely describes a long-established[51] interchange of plants seen so clearly in the history of Earth since the last Ice Age. Designers and urbanists should note this criticism; many of them are now considering novel ecosystems and non-analogue communities as if a new phenomenon. It is not. As many botanists and palaeobotanists have pointed out 'no explicit, irreversible ecological thresholds allow distinctions between "novel ecosystems" and "hybrid" or "historic" ones'.[52] There is no 'new normal' as some have claimed; change has been normal. As Lucretius noted, 'time changes the nature of the whole world ... so that what she bore she cannot, but can bear what she did not bear before.'[53]

Shift and dispersal for designing with plants for future climate spaces

Climate space is a term referring to the multidimensional climatic conditions of an area.[54] Figure 2.7 shows a theoretical representation of climate shift in a region based on changing temperature and rainfall regimes. I outline here some theoretical strategies and ideas concerning the creation of climate space through design. I have placed these ideas under the general terms of assisted migration and designed assistance or relocation; as well as habitats, holdouts, and refuges as designed climate spaces, including topographic buffering. I introduce the concept of '*long-tailing*' for design solutions to climate change.

The previous dispersal of plants in response to ancient climate change suggests that coming changes will produce a highly fragmented landscape, with dispersed populations that will not persist, but will reshuffle in a series of (once again) novel ecosystems. There will be global differences, not universals. As just two examples of global differences, all plants that currently require above-timberline conditions in the United States mainland will be eliminated this century,[55] and fragmentation will be more important in the tropics than in the boreal forest.[56]

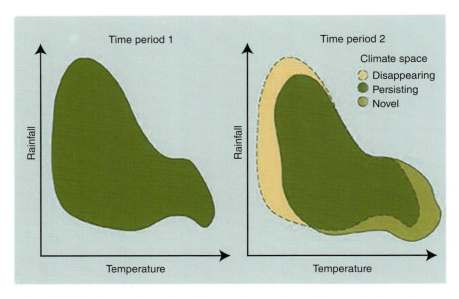

Figure 2.7 This image shows lost, disappearing, and new climates. It gives the two-dimensional climate space (solid outlines) of a region at two time periods. Under changing climatic conditions, species will have to adapt to disappearing and novel climatic conditions at the edges of the region's original climate space. This indicates the increasing importance of the edge, an idea long considered in landscape ecology. Image is from Ralf Ohlemüller, 2011, 'Running out of climate space', *Science* 334: 613–614.

Image: With permission from the American Association for the Advancement of Science.

Our knowledge of the range shifts of the past, conservation studies, and theoretical work are pointing strongly to a change in focus from our historic linking of place and plants to one of populations and dynamics.[57] Nonetheless, plants must live in a place for the duration of their individual lives. For landscape architects, conservationists, and ecologists, the crucial question is how we might assist plants to shift places, perhaps site by site? *Shifting places* is an important idea, because working site by site is what design practice and conservation are about. Can landscape architects assist in the creation of refugia, stepping stones, and holdouts? Which plants will benefit from shifting, and which will stay put? And importantly, should we assist plants to shift places at all?

Individual species and populations will respond over time to coming changes as they did to past changes – wanting to go their separate ways into new ecosystem combinations, or no-analogue groupings. In this manner, plant species will form associations, then the associations will break, and as the species continue to move, the existing associations will dis-associate and new associations will be created, composed of new members, perhaps all entirely new, to yield a wholesale redistribution of species. We can no longer rely on existing species associations, but need to accommodate individual species differences and requirements.[58] Yet this information is almost entirely unavailable, nor do we know the speed of plant movement required or possible to ensure survival.[59] Some species will lag for a few generations, tolerating less than favourable conditions; others might adapt and remain, and others will need to move or will die.[60]

There are a number of important ideas for design and planning about the responses of plant species. First, there will be temporary associations of regional and local species. Second, we will not have assembled these before in plant associations – that is, there have been no modern analogues. Third, quantitative changes in relative proportions of species in mixes will also be changing, just as they did following the last Ice Age.[61] Again, this will not follow globally similar patterns. For example, 10,000 years ago there were similar plant groupings in North America to those found now, but not in Europe.[62]

Creating climate space by migration routes

Many scientists suggest that some intervention is required to ensure tree survival[63] because forecast climate changes are one or two orders of magnitude faster than those achieved during the last glaciation. If a fast migration rate is needed for survival, many tree populations will be tested to the limits of their capacity to adapt and move. Migration capacities of trees are exacerbated by the long life of trees and slow replacement speed compared with species with faster generational renewal.

Intervention is assisted migration, also known as assisted colonisation or managed relocation – assisting plants to shift their ranges in response to climate change. Intervention has become a hotly debated issue in conservation and is particularly focused on those plants that might be restricted in their ability to move. The impetus for this is the knowledge that extinctions rates might be 15–37 per cent by 2050,[64] with a completely 'reshuffled flora' that we saw in the palaeoecological record. There are two types of assisted response. First, actual placement of plants in new areas, that is largely the domain of conservationists; and second, the enhancement of landscape connectivity in which design plays a role.

Designers might engage in exploration of the generation of spatial forms that assist recolonisation.[65] Joshua Lawler and Julian Olden[66] have suggested that helpful strategies would be to determine clear goals of assisted colonisation, ascertain the projected climate impacts, and likely use complementary adaptation strategies, such as enhancing landscape connectivity. Others[67] consider four areas of inquiry can inform the debate over assisted colonisation – forecasting potential climate impacts, gleaning knowledge from previous introductions, looking at the previous palaeohistorical records, and using experiments to measure how a given species might respond in a receiving ecosystem. These ideas point to the need for an extraordinary level of co-operation between science, design, and horticulture, and suggest stronger information exchanges.

Serious ethical issues may limit assisted migration or managed relocations. As various authors have noted, 'It may seem initially an almost moral imperative that humans participate in managed relocations for species truly needing assistance',[68] since current climate change is due to human activity. However, a number of conundrums complicate this apparent moral imperative. All fall into the category – that we might do harm. A major risk is that plant assistance might provide a pathway for biological invasions,[69] and a managed relocation might harm species in the receptor site. We will need to be aware of the risks if, indeed, all risk can be predetermined. Clearly, there will be a spectrum of actions and responses to assistance of plant migration if we can assist at all. Any responses will depend on culture, money, governance, species involved, geography; climate imperative such as water; space availability; politics, planning, and legislative capacity to change quickly.[70] Additionally, the degree of risk acceptance and scientific uncertainty in various regions will vary significantly, and any decision might end as a choice between two evils – to do nothing (that will likely lead to extinction), or to incur a biological penalty with an unanticipated risk and cause disruption and possibly extinction in a receiver landscape.[71] The entire area of assisted migration is murky and uncharted waters. Because it is likely to produce unintended consequences, clear decisions appear to be a long way off.[72] This is an ongoing and complex debate in science.[73]

Creating climate space: Not by connectivity alone

Ideas of connectivity and corridors have long formed the theoretical basis of landscape studies, and designers have been involved in the creation of corridors as linkages for plants and animals. Most notable in this regard has been the influence, since 1996, of *Landscape Ecology Principles in Landscape Architecture and Land-Use Planning*[74] by Wenche Dramstad, James Olson, and Richard Forman. Designers have presented deliberate connectivity responses to climate change. Mary Carol Hunter, an ecologist and landscape architect in Michigan, has suggested ranking plant species for resilience to climate change, to help designers create corridors for assisted migration for plants.[75] This idea is based on connectivity to assist species to shift range in response to climate change in order to survive. Hunter based her suggested rankings on plasticity, functional redundancy, response diversity, and structural diversity. However, data available about habitat needs might restrict this ambition in practice. Indeed, assisted migration requires a highly sophisticated understanding of species biology and ecology, as noted by many conservationists.[76] How will we arrive at this information? Where is it possible to obtain this information?[77] Many mistakes are predicted. This is new territory for everyone concerned. Currently, decisions in this area would have to be made with very incomplete information, not just in science but in the social values tested.[78]

While connectivity is a byword for greening, it is not a cure-all for creating future climate space. Ecological connectivity will be prevented across landscapes in several ways with climate change, urban expansion, and population pressures. Natural green spaces might be too small to allow enough habitat for survival, let alone adaptation, and the pressure on green spaces will be more acute in the coming centuries, whether those spaces are urban, rural, or national parks. Function might also be too small or too slow to be established so that a plant or animal deteriorates beyond the level of functional survival. Barriers from infrastructure are particularly acute in urban areas, and will prevent connectivity. Slowness of legislation can also prevent the creation of connectivity in a timely enough way to be useful, and the legislative changes required to avoid management delays are currently unknown.[79] It is clear from such a simple range of conditions that there will be few global similarities. With such problems of connectivity, the reassurance given by green corridors across plans that appear to link places need to be accompanied by considerable real investments in habitat.

Creating climate space: Designed refugia, stepping stones, and holdouts

In creating climate space, scale is vitally important. Many refuges will be at far finer scales than the regional scale of planning and legislation, and their design requires three-dimensional testing, topographic form generation, and scales that consider various sizes of organisms[80] – this is where design must operate, site by site. Landscape architects will need fine-grained modelling of future movements of plants and plant composition to enable the design of various types of refuges, suggesting acute involvement with ecologists, community, local government, and other landscape scientists.

Species movements will be towards and within holdouts, micro-refugia, and stepping stones.[81] Holdouts are places where species can persist for a long time after their preferred conditions have deteriorated but where they will ultimately become extinct.[82] In contrast to holdouts, micro-refugia are small areas where a species exists despite hostile surroundings; in a micro-refuge a species survives until better conditions return; species are expected to survive in micro-refugia.[83] In the past micro-refugia were important in the preservation of many species and they might be able to play a role in the future. However, we can anticipate that 'favourable' conditions will not return unless we provide them; current climate change is, for the foreseeable future of Earth's history, a very long-term shift. We are, therefore, likely to have many more holdouts (leading to loss) than micro-refugia.

Design might be able to assist holdouts to persist and become longer-lasting full refugia. Persistence is important because it might allow some species to adapt to new local conditions and survive.[84] This entire area of research is absent in science and unexplored in design; it points to a strong need for good collaboration between ecological science and landscape architecture to identify species that might hold out under deteriorating conditions, much as the mammoth did in Wrangel Island in the Arctic Ocean of northern Russia until as recently as 3,700 years ago.[85]

Creating climate space:
Topographic buffering through design

The capacity of landscape architects to test and alter landform is important in the light of recent knowledge of the significance of topographical buffering in climate change survival for many plant and animal species. 'Topographical buffering' refers to the greater climatic stability given by topographically heterogeneous areas than topographically homogeneous areas.[86] Work on micro-refugia by climate change scientist Lee Hannah and his colleagues suggests that topography can boost survival,[87] because folds and shadows, sunnier sides and cooler depths, wind-hollows and exposed areas, all serve to give variation.[88] The clue to potential exploration in design is that during the last Ice Age many species were 'driven by climate, buffered by topography/habitat' for their survival.[89] For example, the mammoths on Wrangel Island survived in topographic niches, where the species adapted by becoming smaller over thousands of years. The key point here is that of providing heterogeneous landscapes at the fine scale.[90,91] Design has the capacity to do this.

Small topographic variations enable topographic buffering that can aid survival of some species (Figures 2.8a and b). Such ideas would need to be tested performatively for both topographic effects and for which species might be able to benefit. Species responses are likely to vary enormously due to differences in their fitness to combat change.[92] 'Topographic buffering' could include the rough and irregular

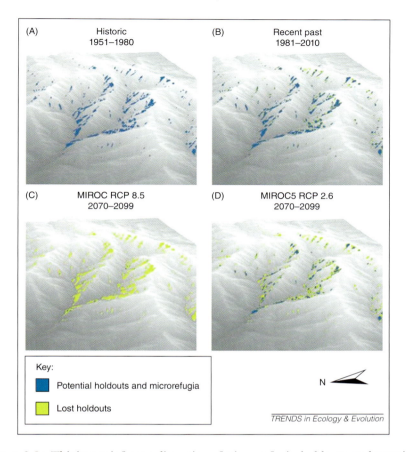

Figure 2.8a This image is from a discussion of micro-refugia, holdouts, and stepping stones by Lee Hannah and his group. The images show potential holdouts and micro-refugia (blue) and loss of holdouts (yellow) across two representative concentration pathways (RCPs: C and D) by the end of the twenty-first century. Here, the mapping is 'based on 30-m statistical downscales of water year climatic water deficit (mm) in the Tehachapi Mountains, California and are overlaid on a 30-m digital elevation model (DEM)'. The authors used water as a surrogate for microclimates likely to be occupied by plant populations. Blue and yellow patches represent relatively wet conditions less than two standard deviations from 1951 to 1980 historical mean. For details see: Hannah *et al.*, 2014, 'Fine-grain modelling of species' response to climate change: holdouts, stepping stones, and micro-refugia', *Trends in Ecology and Evolution* 29(7), 390–397. Of interest to designers is that a heterogeneous landscape could be mountains, or far smaller pieces of topography, such as a wall important for reptiles and insects, a field, or ridges in a park. Topographic buffering and its roughness and opportunities for variation works at all scales, and inside the soil.

Image: Permission from Elsevier.

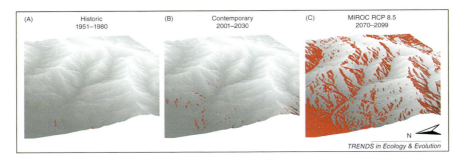

Figure 2.8b Stepping stones can create range shifts and thus climate space for populations of plants and animals, whether at large or fine grains. Here, a species requiring dry conditions is shifting its range. Figure shows stepping stones, moving from historic conditions of 1951–1980 (left) existing conditions (2001–2030), and some predicted conditions (2070–2099). The red are dry areas, and suggest range shifts via stepping stones in California with increasing dryness. The stepping stones might represent a 'leading edge' of movement, or the centre of a new range for that species. Adapted from Hannah, *et al.*, 2014, where the projection was carried out by simulation of future water deficits.

Image: Permission from Elsevier.

topography of the city, which provides heterogeneity of surfaces, shadows, crevices and alley-niches, and temperature. These figures (above) could represent mountains, or little cracks in pavement. We can design at various scales depending on what ambition we have, and what is being assisted.

Techniques that designers now have readily at their disposal through digital computation suggest that we might create – at the small scale in urban, suburban, and even rural areas – topographic variation to assist the creation of climate space. Would small species be assisted by simple topographic manipulations of urban form in streets, neighbourhoods, and larger public parks? Again, this type of manipulation might be more meaningful in some climates and regions than in others. Buffering by topography and constructing ecologies in this manner, with small habitat patches of difference, gives us further warning of avoiding the general approach to ideas of how climate will make an impact in regions, and what our responses might be – whether we are planners, conservationists, or designers.

Creating climate space: Long-tailing the little pieces

Can small patches of climate spaces – such as little patches, voids, and interstitial spaces – assist survival? Small incursions on behalf of biodiversity are important at all the smaller scales of urban and suburban. Here I am suggesting the concept, taken from business, of the long tail; the idea itself is derived from mathematics. In business the long tail expresses the idea that markets are moving away from a small number of mainstream products and markets and into small niches. The reason is due to a simple realisation expressed in the curve in Figure 2.9. This figure shows both the 'long tail' and the 'short head' as they are called in statistics, and, strikingly, they are about the same size. If we think about this in terms of adding biodiversity through landscape design or conservation work, we note that doing lots of little bits might be just as effective as doing the few, very expensive and high-end projects that attract design media attention. The short head is important, but we can also address the long tail, as has been done in business. The example of the long tail usually given is Amazon books; most retailers could not stock on their shelves much beyond the short head, but Amazon has thrived on

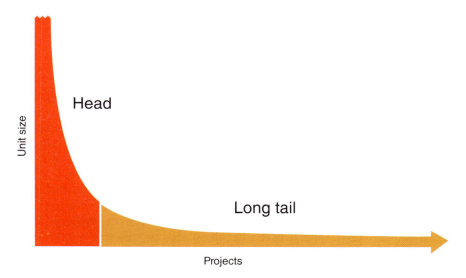

Figure 2.9 The idea of the short head and the long tail from business is here shown for approaches to design. On the left (y) axis, we have cost or unit size of the project and on the bottom (x) axis, the number of projects. For design, the short head can be interpreted as major projects and the long tail smaller incursions for biodiversity, climate survival, topographic heterogeneity, water conservation, soil carbon sequestration, or simple street greening, which together make substantive impacts. Small projects are also more approachable for community groups, developing countries, single operators, and can be done in ready stages, such as greening the streets of one neighbourhood. These projects shift us forward.

Image: Derived from http://www.longtail.com/about.html.

supplying the millions of books in the long tail. Likewise, designers can address the millions of little sites in the long tail of small projects.

The idea of the long tail might also shift a common occurrence in design – the high-profile project that has a lot of publicity but does little of substance in the total scheme of the wider environment, and might exist as an island in a sea of absence. Vast acreages of impermeable surfaces and suburbs could be 'long-tailed' by small incursions at local levels, with combined biodiversity outcomes far exceeding those of the short-tail, high-profile projects. Long-tailing sites could support climate change adaptation, biodiversity preservation, biodiversity assistance, and create climate space. Design has for some time been concerned with voids and interstitial spaces and how these might be treated. Here we might systematically address the long tail as a strategy to add biodiversity to road verges, the permeable footpath, private spaces in suburbia, interstitial spaces ... and all the little bits that we can find. This idea also suggests a strong citizen involvement and citizen science, street by street and house by house.

Global differences in a spectrum of design responses to exotic and native

Discussion of the changing ranges of plant species, how they have shifted and how they might shift again in future, and how we might or might not assist, takes us right into the heart of a major debate in landscape design: what is native, and does it matter? This debate really leads us nowhere if we are looking for a single global answer. Global history tells us that there will never be a single answer. There is no simple solution, no fitted model, but a spectrum of differences. What might these differences be, and where might they lead us in design?

Much of the literature in landscape architecture related to biodiversity appears to take two extreme positions, both deeply unhelpful to thoughtful design. One is that whether native or exotic is a trite issue and still a matter of personal preference, and the second is that preservation of unique flora is a lost cause. Look at all those mixes of weeds that grow happily and 'add' to biodiversity, with the added derision of those 'alien' haters, as if exotic plants relate to human immigrants. These are both erroneous positions difficult to uphold, given knowledge about the macro-ecological history of the world and coming climate changes in the next centuries. In addition, biodiversity loss is a major driver of the loss of performative function in ecosystems.[93]

The design professions are likely unaware that the issue of biological diversity preservation and the rise of mixed non-native and endemic species, notably in our cities, are major issues in biological conservation. There is also the profound ethical issue in that, having mixed up and homogenised the world's plants, we now appear to be saying that it is too hard to repair or maintain, and to 'go with the flow' towards increased homogenisation. The issue for conservation is: 'Have we given up?' And if so, why should we? This discussion does not appear to be occurring with criticality in the design professions. Rather, 'mixtures' of species are often being taken as increased biodiversity, when this is incorrect. Biodiversity is not species numbers alone, as botanists articulated very clearly in the first conferences and definitions of biodiversity. Weeds or exotics rarely 'add' to biodiversity. Biodiversity is more concerned with ecosystem function, a very hard thing to measure, if measurement is possible.[94] Biodiversity is inseparable from the ecosystem at all levels. Indeed, if numbers were taken alone, the native soil biota would define the numbers, and if natives are not used, soil biota will also decline, unseen and unmourned by designers. Their loss in turn will limit the success of other native plants, as they are also part of regional ecological interactions and biodiversity. They are unique and work differently across geographic scales – in other words, all soil microbiota are not found everywhere, and plants interact differently with soil biota dependent on whether they are in their 'home soil' or not.[95] General lack of acknowledgement of the impacts of species planting on root-soil biology is a gamble being played in ignorance by designers.

Perhaps the central problem of thinking about 'native' versus 'exotic' stems from the failure of many designers to see that, while ecological design and constructed ecologies are cultural constructs, all information is not a cultural construct, as noted in Chapter 1. There has been a tendency to see the questions related to native and exotic as a binary of opinion alone. Yet biological facts exist that should not be ignored; facts can assist design and give design discipline. I echo here the comment by the zoologist and evolutionist Richard Dawkins – that some issues are not a matter of opinion but are indisputable. We can note for landscape design such facts as: some plants like shade and others like sun; some plants are intolerant of high phosphate, some 'resent' limestone (as horticulture so lovingly puts it); some plants require pollination by birds that are evolved to feed from pistils of a certain shape; other plants require bees for pollination; some seeds require fire to germinate. There is more genetic diversity in the Australian flower the kangaroo paw (*Anigozanthos manglesii*) in one square kilometre in south-western Australia than in all the spruce (*Picea abies*) in North America.[96] That is a fact of genetics, not a debatable construct of societal thought, or an idea of mere opinion and preference; wishing that it might not be correct will not change it one bit.

Much writing in landscape architecture has failed to address this difference between facts and personal opinion or preference. The landscape architect and theorist James Corner wrote in 1997, 'Terms such as foreign and exotic betray an exclusivity and privileging of the native',[97] without explaining why he considered it to be so. Nor did he note that privileging endemic or the locally endemic – with the term 'native' far more widely understood by the public – might be an excellent idea ecologically. It is important that we learn just why some places might need to have local species supported by us; this is making decisions based on sound knowledge. Consideration of site differences is more productive than complaints against 'scientific ecology' – presumably ecological science,[98] which gives us excellent knowledge of species and their part in local food webs. For example, the botanists Stephen Hopper and Paul Gioia discuss at length local endemism in both flora and fauna in one of the world's great biodiversity hotspots, the Southwest Australian Floristic Region.[99] There, and typical of islands, 40 per cent of all plants are bird-pollinated. In Europe, none are. What then, might be the impacts of our planting decisions in both regions? Clearly, we should give thought on the actions we are taking in each particular region. Global differences exist.

The question of what is exotic is perhaps more clearly answered in the Southern Hemisphere, and the entire question of planting exotics is a more acute part of the spectrum of importance in the Southern than Northern Hemisphere. The isolation of the Southern floras from the North and from each other has been substantive, and there are great differences in plant genera. While most exotics planted in the Northern Hemisphere are from other parts of the Northern Hemisphere,[100] most exotics planted in the Southern Hemisphere are from the Northern.[101,102] As a result, exotics in the Northern Hemisphere are more likely to have kissing cousins of generally small genetic differences and still provide wildlife support, while in the Southern Hemisphere exotics have been described as 'friendless strangers',[103] giving little wildlife or other ecological support due to large evolutionary differences from the plant species and animals that now live with them. As an example, in south-western Australia a small study by insect ecologist D. Clyne on the populations of insects on London plane trees (*Platanus orientalis*) versus those on the local marri tree *Eucalyptus calophylla*[104] found that London plane trees carried few insects, while the marri supported massive numbers and a varied array of species.[105] Such a difference in insect numbers will give a commensurate level of support for birds, reptiles, and likely soil biota as well.

Largely unheralded differences between native and exotic also occur. Colour changes. Yellow deciduous trees dominate Northern Europe in autumn, while red dominates North America and East Asia.[106] In Australia, 'native' greens of the dominant *Eucalyptus* and *Banksia* species are different in hue and chroma to those of introduced genera, which are largely from Europe and eastern Asia; Australian leaves are greyer and less bright.[107] A strong greyness will not be a surprise to Australians because poets celebrate the 'grey-green foliage' as a feature of the entire continent, in the same way that Fall is famous in the eastern USA for its red maples.[108]

The seeking of global commonalities concerning exotic and native runs counter to environmental histories, ecological knowledge, and the importance in the design professions of being 'grounded in place'. In 1997 the American landscape architect and theorist Elizabeth Meyer championed the concept of being grounded in place.[109] It is a tenet of landscape design and works well as a joining of culture, ecology, and the differences that we see and can celebrate within human cultures and regional and local ecologies. This joining is the 'culturally animate ecology' that James Corner lamented to be missing from design in 1997.

There is a strong need to recognise that our concerns about what we plant, whether native or exotic, will be on a *spectrum of importance*, and genetics and plant biodiversity will reveal where on the spectrum a region belongs. The lack of appreciation of this spectrum has led to particularly uninformed debate in the design world concerning species plantings.[110] It perhaps reflects a continuing unease about data in many designers, as if data might somehow pollute or obstruct that beast 'creativity'.

It is well to reflect on that great historian of the city, Lewis Mumford, and his comments in the introduction to Ian McHarg's book *Design with Nature* – of the need for 'human cooperation and biological partnership'. Mumford said we need to seek, 'not arbitrarily to impose design, but to use to the fullest the potentialities—and with them, necessarily, the restrictive conditions—that nature offers'.[111] Geographic histories outlined in this chapter, and the resulting genetic differences, suggest major 'restrictive conditions' to guide many aspects of our thinking and designing.

Conclusions

During the writing of this book I neglected my small and run-down 100-year-old garden, someone else's garden of grass and deciduous trees – *Robinia* from the USA, ornamental pear from Europe, olive from southern Europe planted by migrants from Sparta after the Second World War, a quince from south-western Asia, a tree tomato (or tamarillo) from South America, old-fashioned aromatic roses from Persia, and in the grass lawn broadleaf plantain from Europe, and gazanias from southern Africa. Something else emerged from the planetary garden – windmill grass, a local native that had survived the onslaught of the exotics. *Chloris*[112] *truncata* was somehow resilient and spoke of a past existing still with a changed city. I am now working to give *Chloris* space to flourish on a little grass-bank.

New knowledge of past climates and plant changes has put us into a stronger sense of continuity with the past. This is a different position to the alarm of climate change because, no matter what the cause, we are now in a more profound contact with the experiences of both planetary and human history – that of almost continual change, or *shifting continuities*. There is no disjunction with the past, or with the future, we are in our normal state of flux, with differences at hemispheric and local scales for which we need to design. We are now aware of change and can consider, prepare, adapt, and construct on a *spectrum of responses*.

Our new view has displaced the stable world of place, climate, and vegetation seen in classic texts on biomes and their species. We now need to refocus the lens of landscape planting, and refine planting in response to local, regional, and hemispheric differences, with no presumptions of universal treatments. Mindful of the value of *restrictive conditions* that improve design and thinking, we can sharpen landscape planting. Our core ambition is to plan and design climate space for the future generations of species with whom we live. I have suggested here five conceptual ideas to approach this ambition; there will be more. As climate spaces I outlined the creation of migration routes, the need to address problems hindering connectivity, the creation through research and design testing potential refugia and micro-refugia, the exploration of topographic buffering and heterogeneity, which might be in the canyons of the city and the spaces of suburbia, and the importance of long-tailing designs for more widespread greening for future generations.

Notes

1 *Sphinx and the Pyramids of Ghiza by Facchinell, BNF Gallica*, by Beniamino Facchinelli (1829?–1895?), Bibliothèque nationale de France. Licensed under Public Domain via Wikimedia Commons, http://commons.wikimedia.org/wiki/File:Sphinx-and-the-Pyramids-of-Ghiza-by-Facchinelli,-BNF-Gallica.png.
2 Godfrey M. Hewitt, 1999, 'Post-glacial re-colonization of European biota', *Biological Journal of the Linnean Society* 68(1–2): 87–112. Hewitt also cites H.J.B. Birks, 1989, 'Holocene isochrone maps and patterns of tree-spreading in the British Isles', *Journal of Biogeography* 18: 103–115. Beech did not get to Ireland at all.
3 Alexander von Humboldt, 2012, *Views of the Cordilleras and Monuments of the Indigenous Peoples of the Americas: A Critical Edition*, ed. Vera M. Kutzinski and Ottmar Ette, Chicago, IL: University of Chicago Press, p. 148.
4 Phytogeography is concerned with the geographic distribution of plants.
5 See Margaret B. Davis, 1983, 'Quaternary history of deciduous forests of eastern North America and Europe', *Annals of the Missouri Botanic Garden* 70: 550–563.

6 The Urals in Russia run north–south and denote the 'border' of Europe and Asia. The Kjolen Mountains in Norway and Sweden also run north–south; they were completely covered in glaciers in the last Ice Ages. The Apennines in Italy also run north–south but are south of the Alps.
7 Hewitt, 1999.
8 In Sweden, a specimen of Norway spruce, *Picea abies*, has been found to be 9,500 years old; it has vegetatively reproduced as each generation has died. It is known as Old Tjikko, and there is a group of very old trees around the site.
9 W.D. Sellers, 1965, *Physical Climatology*, Chicago: Chicago University Press; maps are given in Donald Rapp, 2012, 'Life and climate in an Ice Age', in *Ice Ages and Interglacials: Measurement, Interpretation, and Models*, Berlin: Springer Praxis Books, pp. 1–15.
10 V.R. Squires, 1988, 'Landscape: A southern hemisphere perspective', *Earth-Science Reviews* 25: 481–484.
11 John R. Stewart, Adrian M. Lister, Ian Barnes, and Love Dalén, 2010, 'Refugia revisited: Individualistic responses of species in space and time', *Proceedings of the Royal Society B* 277: 661–671.
12 Gunnar Keppel, Karel Mokany, Grant W. Wardell-Johnson, Ben L. Phillips, Justin A. Welbergen, and April E. Reside, 2015, 'The capacity of refugia for conservation planning under climate change', *Frontiers in Ecology and the Environment* 13: 106–112.
13 Winter temperature in southern Europe went down to minus 30°C and summer temperature maxima were 15–20°C. See K.J. Willis, 1996, 'Where did all the flowers go? The fate of temperate European flora during glacial periods', *Endeavour* 20(3): 110–114.
14 Hewitt, 1999. Brian Huntley, 1990, 'European vegetation history: Palaeovegetation maps from pollen data – 13 000 yr BP to present', *Journal of Quaternary Science* 5(2): 103–122. Around 13,000 BP the pollen maps of Europe show that plant and tree species spread much more quickly up the east of Europe between the Caspian Sea and White Sea (35°E) than in the central and western parts. See also: Brian Huntley and H.J.B. Birks, 1983, *An Atlas of Past and Present Pollen Maps for Europe, 0–13,000 Years Ago. no. pt. 2*, available online at doi:10.1016/0034-6667(86)90044-8.
15 Stewart, Lister, Barnes, and Dalén, 2010.
16 P.C. Tzedakis, I.T. Lawson, M.R. Frogley, G.M. Hewitt, and R.C. Preece, 2002, 'Buffered tree population changes in a Quaternary refugium: Evolutionary implications', *Science* 297: 2044–2047.
17 Brian Huntley, 1990, 'European vegetation history: Paleovegetation maps from pollen data – 13,000BP to present', *Journal of Quaternary Science* 5: 103–122.
18 Hewitt, 1999.
19 Brian Huntley, 1991, 'How plants respond to climate change: Migration rates, individualism and the consequences for plant communities', *Annals of Botany* 67 (Supplement 1): 15–22.
20 Brian Huntley discussed many aspects of the changing ecologies of the period following the retreat of the ice back to the poles in a series of papers in the 1990s.
21 The botanist and plant morphologist Edmund Sinnott first linked generation-time and evolutionary rate in E.W. Sinnott, 1916, 'Comparative rapidity of evolution in various plant types', *American Naturalist* 50: 466–478. Rémy Petit and Arndt Hampe note that this work was overlooked when genetic and evolutionary work began to be examined later in the twentieth century. R.J. Petit and A. Hampe, 2006, 'Some evolutionary consequences of being a tree', *Annual Review of Ecology and Systematics* 37: 187–214.

22 Antoine Kremer, Valérie Le Corre, Rémy J. Petit, and Alexis Ducousso, 2010, 'Historical and contemporary dynamics of adaptive differentiation in European oaks', in DeWoody, *et al.* (eds), *Molecular Approaches in Natural Resource Conservation and Management*, Cambridge: Cambridge University Press, pp. 101–122.
23 Donatella Magri, Giovanni G. Vendramin, Bernard Comps, Isabelle Dupanloup, Thomas Geburek, *et al.*, 2006, 'A new scenario for the Quaternary history of European beech populations: Paleobotanical evidence and genetic consequences', *New Phytologist* 171(1): 199–221.
24 Huntley and Birks, 'An Atlas of Past and Present Pollen Maps for Europe, 0–13,000 Years Ago'.
25 Hewitt, 1999.
26 Ibid., Magri, *et al.*, 2006, detail the manner in which *Fagus* moved and stopped across the various regions of Europe, and note the work done on American flora. Their macro-fossil work suggests that 'small populations of beech were unable to increase their role in forest communities for thousands of years'.
27 L. Dobrovolný and V. Tesar, 2010, 'Extent and distribution of beech (*Fagus sylvatica* L.) regeneration by adult trees individually dispersed over a spruce monoculture', *Journal of Forest Science* 56(12): 589–599. For mechanism of seed dispersal of beech see: F.J. Turček, 1961, 'Ecological relationships of birds and woody plants', in *Ökologische Beziehungen der Vögel und Gehölze*, Bratislava: SAV, p. 329.
28 Rémy J. Petit, Emmanuel Pineau, Bridgette Demesure, Roberto Bacilieri, Alexis Ducousso, and Antoine Kremer, 1997, 'Chloroplast DNA footprints of postglacial recolonization by oaks', *Proceedings of the National Academy of Science USA* 94: 9996–10001.
29 Rémy J. Petit, *et al.* (27 others), 2002, 'Identification of refugia and post-glacial colonization routes of European white oaks based on chloroplast DNA and fossil pollen record', *Forest Ecology and Management* 156: 49–74. This work shows the current position of white oaks.
30 Huntley, 1991.
31 This is the same issue that modern crops face, which I will mention in Chapter 4. Modern crops are vast monocultures with little genetic variability.
32 Travis Beck, 2013, *Principles of Ecological Landscape Design*, Washington, D.C.: Island Press, pp. 29–30, notes that plant distribution has been impacted by the last Ice Ages, but does not note any potential inspirations for design.
33 See: Lindsey Gillson and Rob Marchant, 2014, 'From myopia to clarity: Sharpening the focus of ecosystem management through the lens of palaeoecology', *Trends in Ecology and Evolution* 29(6): 317–325. They note that a temporal perspective is needed for the management and restoration of ecosystems that are variable and fluxing. They suggest that palaeoecology, environmental history, along with satellite and other data, can assist scenario building.
34 Hewitt, 1999.
35 For southern Africa, see: Peter Goldblatt and John C. Manning, 2002, 'Plant diversity of the Cape Region of southern Africa', *Annals of the Missouri Botanical Garden* 89(2): 281–302; for SW Australia see: Stephen D. Hopper and Paul Gioia, 2004, 'The Southwest Australian Floristic Region: Evolution and conservation of a global hot spot of biodiversity', *Annual Review of Ecology, Evolution, and Systematics* 35: 623–650.
36 Paul D. Henne, Che Elkin, Jorg Franke, Daniele Colombaroli, Camilla Calo, *et al.*, 2015, 'Reviving extinct Mediterranean forest communities may

improve ecosystem potential in a warmer climate', *Frontiers in Ecology and Evolution* 13(7): 356–362.
37 Elizabeth Cooper, at the Arctic University of Norway, writes of changes in the Arctic in 'Warmer shorter winters disrupt Arctic terrestrial ecosystems', *Annual Review of Ecology, Evolution, and Systematics* 45: 271–295.
38 L. Boisvert-Marsh, C. Périé, and S. de Blois, 2014, 'Shifting with climate? Evidence for recent changes in tree species distribution at high latitudes', *Ecosphere* 5(7): 83.
39 For an excellent review of this subject see: S. Gauthier, P. Bernier, T. Kuuluvainen, A.Z. Shvidenko, and D.G. Schepaschenko, 2015, 'Boreal forest health and global change', *Science* 349: 819–822.
40 Yu Liang, Hong S. He, ZhiWei Wu, and Jian Wang, 2014, 'Effects of environmental heterogeneity on predictions of tree species' abundance in response to climate warming', *Environmental Modelling and Software* 59: 222–231.
41 Jacqueline E. Mohan, Roger M. Cox, and Louis R. Iverson, 2009, 'Composition and carbon dynamics of forests in northeastern North America in a future, warmer world', *Canadian Journal of Forest Research* 39: 213–230.
42 It is fascinating to see that moving north with spruce are grizzly bears. Due to this animal migration, polar bears (*Ursus maritimus*) and grizzly bears (*Ursus arctos horribilis*) are mating. This is not surprising, given that the polar bear evolved from grizzlies about 200,000–300,000 years ago and that now their paths are increasingly intersecting with warmer weather. Resultant offspring have been given the cheeky name of grolar bears. The relationship is discussed in S.L. Talbot and G.F. Shields, 1996, 'Phylogeography of brown bears (*Ursus arctos*) of Alaska and paraphyly within the Ursidae', *Molecular Phylogenetics and Evolution* 5: 477–494. See also Laurence C. Smith, 2011, '*The New North: The World in 2050*, London: Profile Books, p. 3.
43 Ryan K. Danby and David S. Hik, 2007, 'Variability, contingency and rapid change in recent subarctic alpine tree line dynamics', *Journal of Ecology* 95(2): 352–363.
44 This term, the death of future continuities, is from the anthropologist Deborah Bird Rose.
45 John A. Wiens, Nathaniel E. Seavy, and Dennis Jongsomjit, 2011, 'Protected areas in climate space: What will the future bring?', *Biological Conservation* 144: 2119–2125.
46 For a discussion of how reserves are currently designated and their future needs, see: Miguel B. Araújo, Mar Cabeza, Wilfried Thuiller, Lee Hannah and Paul H. Williams, 2004, 'Would climate change drive species out of reserves? An assessment of existing reserve-selection methods', *Global Change Biology* 10(9): 1618–1626.
47 Paul Williams, Lee Hannah, Sandy Andelman, Guy Midgley, Miguel Araújo, *et al.*, 2004, 'Planning for climate change: Identifying minimum-dispersal corridors for the Cape Proteaceae', *Conservation Biology* 19(4): 1063–1074. They say that dispersal is uncertain therefore the best idea is to identify persistence areas that remain suitable for species over time, despite climate change.
48 Richard Weller, 2015, 'World Park', *LA+ Interdisciplinary Journal of Landscape Architecture: Wild*: 10–19.
49 This lovely, but telling, expression comes from Heather M. Kharouba and Jeremy T. Kerr, 2010, 'Just passing through: Global change and the conservation of biodiversity in protected areas', *Biological Conservation* 143: 1094–1101.
50 The issue of the shifting definition of novel ecosystems is discussed critically by Carolina Murcia, James Aronson, Gustavo H. Kattan, David Moreno-Mateos,

Kingsley Dixon, and Daniel Simberloff, 2014, 'A critique of the "novel ecosystem concept"', *Trends in Ecology & Evolution* 29(10): 548–553. These noted ecologists consider that there has been a lack of rigorous scrutiny of the idea and write that, for restoration ecology, there are no irreversible ecological thresholds which have been crossed; though socio-economic and political limitations might have been crossed, these are distinct. The idea of novel ecosystems was first put forward by Fred S. Chapin III and A.M. Starfield, 1997, 'Time lags and novel ecosystems in response to transient climatic change in arctic Alaska', *Climate Change* 35: 449–461.

51 Murcia, *et al.*, 2014, 'A critique of the "novel ecosystem" concept', *Trends in Ecology and Evolution* 29(10): 548–553.
52 Ibid., abstract.
53 As in the front of this book: Lucretius *De rerum natura* 5: 821–836.
54 Ralf Ohlemüller, 2011, 'Running out of climate space', *Science* 334: 613–614.
55 John Grace, Frank Berninger, and Laszlo Nagy, 2002, 'Impacts of climate change on the tree line', *Annals of Botany* 90(4): 537–544; John W. Williams, Stephen T. Jackson, and John E. Kutzbach, 2007, 'Projected distributions of novel and disappearing climates by 2100 AD', *Proceedings of the National Academy of Sciences USA* 104(14): 5738–5742.
56 S. Joseph Wright, 2005, 'Tropical forests in a changing environment', *Trends in Ecology & Evolution* 20(10), 553–560.
57 Lee Hannah, Lorraine Flint, Alexandra D. Syphard, Max A. Moritz, Lauren B. Buckley, and Ian M. McCullough, 2015, 'Place and process in conservation planning for climate change: A reply to Keppel and Wardell-Johnson', *Trends in Ecology and Evolution* 30(5): 234–235.
58 Important here will be the physiological capacities of species to cope with new conditions. A recent book which considers this issue is Michael Tausz and Nancy Grulke, 2014, *Trees in a Changing Environment: Ecophysiology, Adaptation, and Future Survival*, Dordrecht: Springer Science +Business Media.
59 A key to species loss is the speed of plant migration required to avoid extinction. Plant dispersal rates after the last Ice Age that were calculated solely from pollen records are considered to overestimate the ability of most species to respond to rapid climate change in the coming centuries. See Rebecca S. Snell and Sharon A. Cowling, 2015, 'Consideration of dispersal processes and northern refugia can improve our understanding of past plant migration rates in North America', *Journal of Biogeography* 42(9): 1677–1688. Importantly, changes and plant responses will occur at different scales and rates, with some change quite sudden, as has happened in the past. For example, about 11,500 years ago, at the end of a period known as the Younger Dryas, the temperature in Greenland rose by 10°C in a decade. See: Kurt M. Cuffey and Gary D. Clow, 1997, 'Temperature, accumulation, and ice sheet elevation in central Greenland through the last deglacial transition', *Journal of Geophysical Research* 102: 26383–26396.
60 See Sally N. Aitken, Sam Yeaman, Jason A. Holliday, Tongli Wang, and Sierra Curtis-McLane, 2008, 'Adaptation, migration or extirpation: Climate change outcomes for tree populations', *Evolutionary Applications* 1(1): 95–111, for a discussion about the three possible fates of our major tree species.
61 Brian Huntley, 1990, 'European post-glacial forests: Compositional changes in response to climatic change', *Journal of Vegetation Science* 1(4): 507–518.
62 Ibid. Huntley notes that in eastern North America most post-glacial pollen spectra had modern analogues, whereas in Europe many quite recent post-glacial pollen spectra lack modern analogues.

63 See Aitken, Yeaman, Holliday, Wang, and Curtis-McLane, 2008; and Antoine Kremer, 2010, 'Evolutionary responses of European oaks to climate change', *Irish Forestry*, online pdf, 53–65. Genetic measures are suggested in Kremer. See also Scott R. Loarie, Phillip B. Duffy, Healy Hamilton, Gregory P. Asner, Christopher B. Field, and David D. Ackerly, 2009, 'The velocity of climate change', *Nature* 462: 1052–1055; these authors consider that the climates of only 8 per cent of global protected areas have residence times exceeding 100 years.

64 This figure is for Europe; W. Thuiller, S. Lavorel, M.B. Araújo, *et al.*, 2005, 'Climate change threats to plant biodiversity in Europe', *Proceedings of the National Academy of Science USA* 102: 8245–8250.

65 Richard T. Corlett argued that the prefix 're' means all associations go back to some previous condition. However, this is not the case as 're' can also mean 'again', as in 'reorganise'. See: R.T. Corlett, 2016, 'Restoration, reintroduction, and rewilding in a changing World', *Trends in Ecology and Evolution* 31(6): 453–462.

66 This discussion is based on the work articulated by Joshua J. Lawler and Julian Olden, 2011, 'Reframing the debate over assisted colonization', *Frontiers in Ecology and the Environment* 9(10): 569–574.

67 Ibid.

68 Sarah Reichard, Hong Liu, and Chad Husby, 2012, 'Is managed relocation of rare plants another pathway for biological invasions?', in Joyce Maschinski and Kristin E. Haskins (eds), *Plant Reintroduction in a Changing Climate: Promises and Perils*, Washington, D.C.: Island Press, pp. 243–261.

69 Ibid., p. 244.

70 A summary of ecological and social considerations for assisted relocation appears in David Richardson, *et al.* (21 others), 2009, 'Multidimensional evaluation of managed relocation', *Proceedings of the National Academy of Sciences USA* 106(24): 9721–9724.

71 Marko Ahteensuu and Susanna Lehvävirta, 2014, 'Assisted migration, risks and scientific uncertainty, and ethics: A comment on Albrecht *et al.*'s review paper', *Journal of Agricultural and Environmental Ethics* 27: 471–477.

72 Anthony Ricciardi and Daniel Simberloff, 2009, 'Assisted colonization is not a viable conservation strategy', *Trends in Ecology and Evolution* 24(5): 248–253.

73 N. Hewitt, N. Klenk, A.L. Smith, D.R. Bazely, N. Yan, *et al.*, 2011, 'Taking stock of the assisted migration debate', *Biological Conservation* 144(11): 2560–2572. Mark W. Shwartz discusses assisted migration and notes that thermal distance is critical. He notes three critical challenges: understanding the biology, evaluating biological risks, and negotiating social discord and concern; see Shwartz, 2016, 'Elucidating biological opportunities and constraints on assisted colonization', *Applied Vegetation Science* 19(2): 185–186.

74 Wenche E. Dramstad, James D. Olson, and Richard T.T. Forman, 1996, *Landscape Ecology Principles in Landscape Architecture and Land-Use Planning*, Washington, D.C.: Harvard University Graduate School of Design, Island Press, & American Society of Landscape Architects.

75 Mary Carol Hunter, 2011, 'Using ecological theory to guide urban planting design: An adaptation strategy for climate change', *Landscape Journal* 30(2): 173–193.

76 See the discussion in: Joyce Maschinski, Konald A. Falk, Samuel J. Wright, Jennifer Possley, Julissa Roncal, and Kristie S. Wendelberger, 2012, 'Optimal locations for plant reintroductions in a changing world', in Joyce Maschinski and Kristin E. Haskins (eds), *Plant Reintroduction in a Changing Climate: Promises and Perils*, Washington, D.C.: Island Press, pp. 109–129. Also the

erude paper by David Richardson, *et al.*, 2009, 'Multidimensional evaluation of managed relocation', *Proceedings of the National Academy of Sciences USA* 106(24): 9721–9724.

77 Andrew J. Davis, Linda S. Jenkinson, John H. Lawton, Bryan Shorrocks, and Simon Wood, 1998, 'Making mistakes when predicting shifts in species range in response to global warming', *Nature* 391: 783–786. The authors discuss the whole idea of climate envelopes and discuss the problem of changed relationships between species.

78 David Richardson, *et al.*, 2009.

79 G. Hole, B. Huntley, J. Arinaitwe, S.H.M. Butchart, Y.C. Collingham, L.D.C. Fishpool, D.J. Pain, and S.G. Willis, 2011, 'Towards a management framework for key biodiversity networks in the face of climatic change', *Conservation Biology* 25(2): 305–315.

80 For a discussion about the size of organisms and climate, and the current spatial mismatch between the size of organisms and the scale at which climate data is collected, see: Kristen A. Potter, H. Arthur Woods, and Sylvain Pincebourde, 2013, 'Microclimatic challenges in global change biology', *Global Change Biology* 19: 2932–2939.

81 Lee Hannah, Lorraine Flint, Alexandra D. Syphar, Max A. Moritz, Lauren B. Buckley and Ian M. McCullough, 2014, 'Fine-grain modeling of species' response to climate change: Holdouts, stepping-stones, and microrefugia', *Trends in Ecology and Evolution* 29(7): 390–397.

82 Kristoffer Hylander, Johan Ehrlén, Miska Luoto, and Eric Meineri, 2015, 'Microrefugia: Not for everyone', *Ambio* 44: S60–S68.

83 Valentí Rull, 2009, 'Microrefugia', *Journal of Biogeography* 36: 481–484.

84 Mark J. McDonnell and Amy K. Hahs, 2015, 'Adaptation and adaptedness of organisms to urban environments', *Annual Review of Ecology, Evolution and Systematics* 46: 261–80.

85 S.L. Vartanyan, V.E. Garutt, and A.V. Sher, 1993, 'Holocene dwarf mammoths from Wrangel Island in the Siberian Arctic', *Nature* 362: 337–340. The authors attributed survival of a mammoth population on Wrangel Island to local topography and climatic features, which permitted relictual preservation of communities of steppe plants. See also R. Dale Guthrie, 2004, 'Radiocarbon evidence of mid-Holocene mammoths stranded on an Alaskan Bering Sea island', *Nature* 429: 746–749.

86 Nicole E. Heller, Jason Kreitler, David D. Ackerly, Stuart B. Weiss, Amanda Recinos, Ryan Branciforte, Lorraine E. Flint, Alan L. Flint, and Elisabeth Micheli, 2015, 'Targeting climate diversity in conservation planning to build resilience to climate change', *Ecosphere* 6(4): 1–20. These authors examined eighteen climate scenarios for San Francisco Bay, and found that the climate-based network planned at the sub-regional scale captured a greater range of climate space and showed higher climatic stability than the vegetation and regional-based networks.

87 Hannah, Flint, Syphard, Moritz, Buckley and McCullough, 2014.

88 These ideas can also be explored in the canyons of the city and the spaces of suburbia.

89 P.C. Tzedakis, I.T. Lawson, M.R. Frogley, G.M. Hewitt, and R.C. Preece, 2002, 'Buffered tree population changes in a Quaternary refugium: Evolutionary implications 2002', *Science* 297(5589): 2044–2047; and see: Susana Nieto-Sánchez, David Gutiérrez, and Robert J. Wilson, 2015, 'Long-term change and spatial variation in butterfly communities over an elevational gradient: Driven by climate, buffered by habitat', *Diversity and Distributions* 21(8): 950–961.

90 This is discussed in some detail in Kristoffer Hylander, Johan Ehrlén, Miska Luoto, and Eric Meineri, 2015, 'Microrefugia: Not for everyone', *Ambio* 44 (Suppl.): S60–S68.
91 Of note is that the relationship between heterogeneity and species diversity might not be a simple one. Recent work suggests that for some situations there will be a decline in particular species with increased heterogeneity provision. This is known as the area-heterogeneity trade-off; any increase in environmental heterogeneity within a fixed space must lead to a reduction in the average amount of effective area available for individual species. See Omri Allouche, Michael Kalyuhny, Gregorio Moreno-Rueda, Manuel Pizarro, and Ronen Kadmon, 2012, 'Area-heterogeneity tradeoff and the diversity of ecological communities', *Proceedings of the National Academy of Sciences* 43: 17495–17500.
92 C.D. Pigott and Jacqueline P. Huntley, 1981, 'Factors controlling the distribution of *Tilia cordata* at the northern limits of its geographical range. III. Nature and causes of seed sterility', *New Phytologist* 87: 817–839.
93 David U. Hooper, E. Carol Adair, Bradley J. Cardinale, Jarrett E.K. Byrnes, Bruce Hungate, *et al.*, 2012, 'A global synthesis reveals biodiversity loss as a major driver of ecosystem change', *Nature* 486(7401): 105–108.
94 See a discussion by J.P. Grime, 1997, 'Biodiversity and ecosystem function: The debate deepens', *Science* 277(5330): 1260–1261. Although 20 years old, the comments are still pertinent.
95 Marnie E. Rout and Ragan M. Callaway, 2012, 'Interactions between exotic invasive plants and soil microbes in the rhizosphere suggest that "everything is *not* everywhere"', *Annals of Botany* 110: 213–222.
96 Personal comm. Stephen Hopper, Director Royal Botanic Gardens Kew 2006–2012.
97 James Corner, 1997, 'Ecology and landscape as agents of creativity', in George F. Thompson and Frederick R. Steiner (eds), *Ecological Design and Planning*, New York: John Wiley & Sons, pp. 80–108.
98 This is a lot like complaining that music is musical; ecology is a science.
99 Stephen Hopper and Paul Gioia, 2004, 'The Southwest Australian floristic region: Evolution and conservation of a global hot spot of biodiversity', *Annual Review of Ecology, Evolution and Systematics* 35: 623–50.
100 An exception are many species of *Eucalyptus* from the Southern Hemisphere planted in both North Africa and California for their shade or other properties.
101 See M.L. McKinney, 2005, 'Species introduced from nearby sources have a more homogenizing effect than species from distant sources: Evidence from plants and fishes in the USA', *Diversity and Distributions* 11(5): 367–374; F.A. La Sorte, M.L. McKinney, and P. Pyšek, 2007, 'Compositional similarity among urban floras within and across continents: Biogeographical consequences of human-mediated biotic interchange', *Global Change Biology* 13(4): 913–921. For further discussion, see: M.J. Grose, 2016, 'Green leaf colours in a suburban Australian hotspot: Colour differences exist between exotic trees from far afield compared with local species', *Landscape and Urban Planning* 146: 20–28.
102 An exception here are the invasive Australian *Acacia* species in southern Africa (although used for firewood), and veldt grass from southern Africa that flourishes in Australia, as if tit-for-tat.
103 D. Clyne, 1984, *More Wildlife in the Suburbs*, Sydney: Angus and Robertson.
104 Now known as *Corymbia calophylla*.
105 Clyne, 1984.

106 Simcha Lev-Yadun and Jarmo K. Holopainen, 2009, 'Why red-dominated autumn leaves in America and yellow-dominated leaves in Northern Europe?', *New Phytologist* 183: 506–512.
107 Grose, 2016, *Landscape and Urban Planning* 146: 20–28.
108 As pointed out by Simcha Lev-Yadun and his co-workers, colour impacts predation because animals have evolved with plant colour. For fuller discussions see: Simcha Lev-Yadun, M. Inbar, I. Izhaki, G. Nèman, and A. Dafni, 2002, 'Colour patterns in vegetative parts of plants deserve more research attention', *Trends in Plant Sciences* 7(2): 59–60; S. Lev-Yadun, A. Dafni, M.A. Flaishman, M. Inbar, I. Izhaki, G. Katzir, *et al.*, 2004, 'Plant coloration undermines herbivorous insect camouflage', *BioEssays* 26: 1126–1130.
109 Elizabeth K. Meyer, 1997, 'The expanded field of landscape architecture', in George F. Thompson and Frederick R. Steiner (eds), *Ecological Design and Planning*, New York: John Wiley & Sons, pp. 45–79.
110 Margaret Grose, 2012, 'Deepening the thinking in ecology for design', *Landscape Architecture Australia* 130: 30.
111 Lewis Mumford's Introduction in Ian McHarg, 1969, *Design with Nature*; viii.
112 Chloris is the Greek goddess who transformed Narcissus, Crocus, and Hyacinthus into flowers.

3

SHIFTING ADAPTABILITIES, NOT STATIC CONCEPTS

Error veritate simplicior
(Error is simpler than truth)

Nietzsche[1]

When I was a child, a book on the 'History of Man' startled my imagination. Like many before me, I wondered: who were 'we' and what is our history? Years later I walked in the pretty wooded Neander Valley, where 'Neanderthal Man' was discovered in 1856 during a mining operation.[2] Two representations of Neanderthals seen at the site tell us a great deal about human inquiry and imagination, and our capacity to change our minds (Figure 3.1).

On the left is the image of the brutish, dim-witted 'other' Man, club in hand, in the imagination of the nineteenth century and for

Figure 3.1 Images of two Neanderthals in the valley of the Düssel River near Düsseldorf, one from the park and the other in the Neanderthal Museum. Neanderthals inhabited Europe as recently as 30,000 to 24,000 years ago.

Images: By the author.

most of the twentieth. On the right a view from the twenty-first century of a Neanderthal as a fellow human type to us, in his family group, concentrating on the learned and difficult task of sharpening a tool, and dressed for his work. With a shave and contemporary clothing, this Neanderthal man might disappear without comment into nearby Düsseldorf's tram system in the peak-hour rush. It is clear that the world view about another human type has radically changed between these two eras. One cannot view the newer models of the Neanderthals and wonder what the world was like for them, and for other human types.

Despite the leap of understanding and knowledge revealed in the images above, within the field of landscape perception a theory is still used that clings to older ideas of how humans lived, where we lived, and what we might have experienced. This is the savannah theory of landscape perception, which has been a dominant idea taken into landscape architecture. The savannah theory was established in the 1980s and 1990s.[3] Gordon Orians and Judith Heerwagen hypothesised that 'people have a generalised bias towards savanna[h]-like environments'[4] even when they have had no direct experience of that type of landscape. They saw this as an 'evolved response' to landscape due to human evolutionary history, and predicted that 'the response to other types of biomes, such as desert, steppe, and closed forest, require direct experience.'[5] Savannah landscape as experienced by early humans was believed to be comprised of grassland and scattered trees, and this particular landscape type was somehow imprinted upon us as the most desirable, influencing our mental preferences for landscape ever since. Designers have tightly held these ideas of the savannah theory for more than thirty years with little critique. In this chapter in the light of new knowledge I reassess the savannah theory and critique its assumptions of a universal, global, and dominant view of landscape perception.

A plethora of papers have cited the importance of savannah landscape as a permanent perceptual legacy of human history, rather than as the theory it was. I believe that this theory is now unsupportable. Indeed, Orians and Heerwagen suggested that their ideas would be invalidated or refuted because the history of science is one of discard and revision,[6] as seen in the representations of the Neanderthals. Revision of the savannah theory will offer fruitful and plural directions of inquiry.

This revision of the savannah theory of landscape perception started from two questions. First, why is a revision of the savannah theory important in thinking about constructed ecologies? Some people might consider that perception and construction are poles apart – one occurs inside our head and influences our psyche, and the other is what we do with our hands. However, it is this very dichotomy that I want to examine. The second question is: how is rethinking ideas about archaic and more recent humans important for understanding our capacities for constructed ecologies today? What we do reflects how we think, and what we construct reflects our thinking. Can we then change our thoughts, as we need to do in order to address the important issues today of climate change, population pressures, and food production? Yet the savannah theory implies that we are pre-wired for a certain preference. If we are pre-wired in one preference for a specific landscape type, it does venture the question – if we are pre-wired as the savannah theory suggests, do we and would we find change difficult? The savannah theory implies that human perception has been static, locked in one landscape. My interest in the savannah theory grew from my frank disbelief at this notion, given the wide range of landscapes in which we happily live, and given the varied responses we will need to address with coming environmental shifts and challenges. An ability to change with more information or life experiences, and with the knowledge of the life experiences of others, gives us a brighter outlook. Humans learn.

In critiquing the savannah theory, I draw upon new information and ideas found in the sciences and humanities that deal with human evolution. As I set out the variance of human culture and experience, I paint a picture gained from a more complex understanding of human evolution, the landscapes in which we lived, and our responses. In the course of this discussion, many readers may be surprised to discover just how little evidence there is for the savannah theory, and how much readily available knowledge has been ignored. Readers may be further surprised to discover that anthropology had its own savannah theory and discarded it. I suggest new concepts that can form the basis of more fruitful inquiries into human perception and landscape responses. The overriding question that the savannah theory of universal preferences could not address is surely: how did we as a species respond to our previous environments, and how does this response shed light on our capacity to adapt to change now?

Changing worldviews

Here, in this image (Figure 3.2), we have something extraordinary – a group of *Homo sapiens* visiting a site where another human type once lived. That there were several types of humans on Earth until about 30,000 years ago is now well established.[7] Both scientists and laymen have speculated about how 'we' *Homo sapiens* (then) might or might not have responded to other human groups, but we (now) must be mindful of one simple assumption in that regard when we wonder

A BACKGROUND TO DESIGN

Figure 3.2 Tourists doing something that would have been unthinkable just twenty years ago: visiting the site of another human type who interbred with 'us', and contributed towards 'us' as we are today outside Africa. This is the Denisova Cave, Altai Mountains, south-western Siberia, from where fossils of Denisova people were uncovered in 2009.

Image: 'Turist den-peschera' by ČuvaevNikolaj at ru.wikipedia 2010. Licensed under CC BY-SA 3.0 via Wikimedia Commons – http://commons.wikimedia.org/wiki/File:Turist_den-peschera.jpg#/media/.

about these interactions. We usually assume that Neanderthals or Denisovans in eastern Asia, whose genes millions of us carry,[8] were considered to be different by our ancestors. Other human types might not have been seen as different; we do not know when the concept of 'difference' and 'other' might have arisen, or if a meeting with a group of Neanderthals or Denisovans instilled in *Homo sapiens* any greater level of caution than if meeting any group of strangers, particularly groups whose lifestyles were quite similar, as they all were.[9]

Homo sapiens interbred with Neanderthals over a time range of 37,000 to 86,000 years ago,[10] and with the Denisovans in eastern Asia (Figure 3.2); Denisovans mated with an as yet unknown other *Homo*,[11] but not, it seems, with *Homo floriensis*.[12] Our view of Neanderthals is so greatly different to the nineteenth-century view (Figure 3.1) that contemporary writers about human history refer to Neanderthals as 'human', and even refer to our common ancestors with Neanderthals – *Homo heidelbergensis*, for example – as an 'earlier species of human'.[13] This is a massive shift in the view of who 'we' are.

We now see Neanderthals as our cousins, who had developed in their own right, and did not gain their knowledge and technologies from us,[14] as had long been supposed.[15] While being anatomically different from us, they were equally thinking (sapient) humans. They were likely as bright as we were, likely buried their dead with flowers,[16] spoke,[17] wore jewellery,[18] and had a varied diet including a wide array of plants. They utilised many different landscapes.[19] There are strong claims in anthropology today that the artwork found in the sites in France might not be ours (*Homo sapiens*), but Neanderthal.[20] Discoveries of prehistoric paintings found at eleven locations in Spain, including the UNESCO World Heritage sites of Altamira, El Castillo (4,000 years earlier than the cave of Chauvet), and Tito Bustillo, might be Neanderthal, not ours.[21] Evidence now supports the notion that there was no cognitive gap between Neanderthals and 'their contemporary modern humans'.[22]

Ancient *Homo sapiens* interbred with Neanderthals, and millions of present-day *sapiens* whose ancestors shared landscapes with Neanderthal peoples outside Africa have 1–3 per cent of their genome from Neanderthals.[23] Anthropologists now consider that Neanderthals were 'a geographically widespread, culturally sophisticated, behaviorally variable, highly adaptable, and long-lived hominin'.[24]

Such changes in thought about what it is to be human must surely lead us to ask again what forces, what environments, and what material cultures of construction influenced us in our long journey from the early hominins[25] of Africa to who we happen to be today,[26] and where – dispersed across the globe and ecologically dominant. It is clear from studies in anthropology and genetics that ours is a long, varied journey, with multiple players, not a static moment in time or in space.

Savannah theory of landscape perception

Perhaps surprisingly, the savannah theory had started as a habitat preference concept based on Orians' experience as an ornithologist specialising in blackbirds.[27] He presumed that animals first chose a habitat, followed by a patch selection, then a 'home' site selection, followed by a determination of usable things in that chosen environment.[28] Inherent in Orians' analysis is that humans would behave the same as birds, both today and in archaic times. However, while we might now decide on a tree change or sea change at retirement, then choose a house, then the things to do – following the blackbirds' order – it is surely the opposite for archaic humans, who would have stopped for food and water (which attracts game), found shelter, perhaps stayed seasonally, and then remembered the general area for the following year and hope that the site delivered continuing support to their party. Orians' 1980 paper was loosely based on a selection of real estate values, an assumption of permanence, views of European colonists on the Great Plains of the USA,[29] and a conflation of a spatial form of scattered trees on a grassland with 'the savannah', which is a living entity not merely spatial form. It is a precarious assumption to consider only spatial form in landscape analysis.[30] In the original exposé, the English landscape was, without articulation, classed as 'savannah'-like, as were all public parks. Yet how we would get a fright if out strolling in the gentle green of an English landscape we turned to see ourselves being stalked by large predators – this was the reality of archaic humans on an open savannah, for those who lived there or passed through it on seasonal routes. A great many fossil finds that have informed our understanding of human evolution have come from humans predated upon and carried into caves by leopards.[31]

Twelve years later, Orians teamed with the cognitive psychologist Judith Heerwagen to comment about human responses.[32] Surprisingly, however, the savannah theory did not draw upon information then available within anthropology, but assumed a particular prehistoric setting of one savannah type. Open grasslands with scattered groups of wide-canopied trees epitomised by *Acacia tortilis*,[33] the umbrella thorn tree, was taken as *the* early human habitat upon which *Homo sapiens* thrived, and assumed that a long residence upon that landscape determined habitat preferences in us until today. In this concept there are a number of explicit and implicit assumptions. Explicit assumptions are, first, that a savannah consisted of open grassland and scattered trees; second that the spatial form of the savannah had an overriding importance on human experience and third, that one period of time – living on the savannah of open grassland and scattered trees – set us up with a particular mind-set. Implicit assumptions are that the human brain has not changed since remote time, and that the years of experience with many different landscapes and the tools which we have adapted for our use in constructing and dealing with these different landscapes have not influenced our mind or our perception of either ourselves or our landscapes. The savannah theory suggests that we have been incapable of reassessing and reimagining new impacts and experiences that might have altered our perceptions since prehistory.

Of note is that, in 1992, Orians and Heerwagen stated that discarded evolutionary ideas will have played their role in asking questions of how we use landscape. However, despite the authors' own caution on this matter, the savannah theory in landscape perception has become a lazy idea in landscape research – a theory that has not been criticised fully in the light of new ideas. The savannah theory has usually been stated without analysis, without critique of the assumptions outlined above,[34] and without apparent recourse to research in palaeoanthropology that had already doubted the savannah as the prime environment of relevance to human evolution.[35] Such lack of questioning will lead us nowhere as investigators of the human condition in constructing new ecologies.

Any theory dealing with human evolutionary history and experience must have strong ties to the human evolutionary sciences. But lack of attention to the original field of study by researchers in landscape perception and design raises concern at the level of scholarship using the idea of the savannah theory and gives lessons for landscape architecture. There is peril in applying a theory when those using it have little or no foundational background. We will lack the capacity to critique ideas, and we will likely miss new ideas if we do not continually keep in touch with that discipline by some mechanism.[36] A lack of critique might have acute impacts if the theory lies on the other side of the 'two cultures', the arts and sciences,[37] such as exists between design and the human evolutionary sciences. The central danger is that we might be standing still with a theory derived from a field of study that has evolved while we were not looking.

In the last twenty years our views of Neanderthals have radically altered along with ideas and theories on human evolution. Since the 1980s many new discoveries have been made in the field of anthropology and other disciplines that have been assisting anthropologists tease out who we were. Many other ideas influence how we regard the importance of savannah on human experience. Increasing evidence from anthropology, cognitive archaeology, genetics, and neurobiology suggest more complex evolutionary histories and less simple narratives than that suggested earlier; indeed, there is no linear narrative from prehistory until now.

Landscape architects and other design disciplines also need to deal with the question who to consult for information, and how do we tease out questions when answers sit in another discipline? With the idea of the savannah, Orians clearly went to behavioural psychology to see what legacies a savannah might have left upon our thinking. Palaeoecologists, palaeoanthropologists, and behavioural anthropologists studying primates would have given him a different understanding of what a savannah can be, and how these disciplines regard the impact of the savannah on human evolution and cognition. In seeking out one discipline only, the questions were skewed in a particular direction – straight into perception, and this skew has directed landscape theory ever since. In many ways it was a particularly curious move because a 'savannah theory' already existed within evolutionary theory in anthropology, which had, by 1992, discarded it as being simplistic.

The now-abandoned savannah theory of anthropology suggested that when early hominins entered a savannah lifestyle, the spatial nature of the savannah led to an increase in bipedalism – how we came to walk upright – because we needed to look over the grassland for predators and opportunities. Raymond Dart, who worked on the early hominins in eastern Africa in the 1920s, put forward the role of the African savannah as shaping human evolution.[38] His simple idea caught the public's imagination[39] and became entrenched. That early theory was important because it was the very first key environmental theory that attempted to explain bipedalism.[40] However, by the 1990s it was largely superseded with the 'aridity hypothesis', primarily because the savannah idea neglected to investigate crucial biological aspects of early hominin evolution.[41] The savannah as an important influence on bipedalism has now been questioned by anthropologists as to being of any substance at all.[42] The more complex and complete picture of hominin evolution has emerged as one of co-evolution of humans with their environments,[43] based on their cognitive responses to continual climatic changes in eastern Africa and elsewhere in the last 200,000 years.

Co-evolution with environmental changes suggests movement, flux, and new responses. It is now clear that early human types and early *Homo sapiens* moved seasonally, stayed and rode out climatic changes with their associated vegetation changes and herd movements, or moved in response to those changes. We occupied many landscapes from cold steppe to warm forest.[44] Such different environmental settings suggest resilience and adaptability, not preference for one landscape type. In this journey of human evolution, what were the implications to us as a species of this moving between landscapes? How did it influence our mind, and how might the act of moving into new areas have influenced our thoughts? Did the stresses of change enable us to think differently? How might they have led us to construct landscapes?'

Articulating questions about landscape's savannah theory

There are fundamental questions that we need to ask concerning who we, *Homo sapiens*, are, what landscapes we lived in historically, and how these landscapes might have impacted our thinking. I ask lots of questions. What is savannah? Was it only on savannah that we lived? Who were 'we'? Were there other environments that we inhabited for tens of thousands of years? What of the impacts of environments in which we migrated and settled? Have we not changed since archaic *Homo* and early *Homo* forms and, if we have, how have we? Has our brain changed? Can we ever know what our ancestors thought and preferred long ago? Are there lessons for us as to how we might change again to cope with major environmental challenges in the coming centuries? Evidence from other disciplines provides a more complex reading of prehistory and may answer these questions.

What is savannah? Is it what we have presumed?

What is savannah? Surprisingly for a 'savannah theory' that has been adopted so strongly, virtually no one interrogated the physical savannah and what is meant by that term. It is therefore important to tease out what a savannah might be, and what it is not necessarily.

Savannahs make up more than 20 per cent of the Earth's land surface,[45] suggesting a wide terminology[46] and widely varying environments, both spatially and ecologically. Savannahs support most of the world's livestock and wild herbivores. African savannahs are principally grass and shrub savannahs; tree and shrub savannahs; woodland savannahs; and forest savannah mosaics; and make up about 50 per cent of the total land area of the continent, with the type of savannah largely dependent on rainfall[47] and soils. The Oxford ecologist Frank White described woodland savannah as trees of at least 8m tall and 40 per cent canopy cover, and forest savannah as trees of at least 10m tall and continuous cover,[48] showing that the savannahs cross an enormous range of spatial types, both vertically and horizontally. Woodland savannah is dominated by the *miombo* woodland savannah, named after the Muuyombo tree, *Brachystegia bodhmii*,[49] which is 15–20m tall. At the high-rainfall end of the spectrum of savannah types, the heavily wooded mesic savannahs 'begin to structurally resemble forests'.[50] It is clear that the concept of savannah theory in both twentieth century anthropology and landscape studies refers to one particular type – extensive grassland with scattered trees; that is, neither woodland nor forest. Yet woodland and forest types of savannah are extensive and were extensive during early human evolution. To give an idea of their visual range, Figures 3.3 and 3.4 show a variety of savannahs in Tanzania, all found within a small region.

All the images of savannah reveal wide variation in what a savannah is like. It is clear that savannahs are mosaic environments, and there is a strong seasonal difference between the wet and dry seasons. In the dry season some deciduous trees shed some or all of their leaves, other plants shed leaves in the wet season, and many do not shed their leaves

Figure 3.3 All these images show modern savannah in Tanzania. Alex Piel took these photos in wet and dry seasons.

Images: Alex Piel and Fiona Stewart, Ugalla Primate Project, Tanzania, with permission. See: http://ugallaprimateproject.com/media/photos.

at all. These views are very different from public open spaces in the developed world that are often taken to represent a preference for the spatial form of the African savannah. There appears to be a common misunderstanding that all savannahs look like the Serengeti of Kenya and Tanzania, which has been referred to as the 'classical savannah'.[51] The Serengeti Plains is a vast, largely treeless 'endless'[52] area, yet they

Figure 3.4 Top and lower: Mio woodland savannah, western Tanzania, which are similar in their range to those experienced by early *Homo* species. Both of these images show savannah landscapes, perhaps revealing unexpected conditions for many readers. http://ugallaprimateproject.com/projects/chimpanzee-behavioral-ecology.

Images: Alex Piel and Fiona Stewart, Ugalla Primate Project.

A BACKGROUND TO DESIGN

are also interspersed with hybrid habitats and 'islands in the Serengeti' of rocky outcrops and dense vegetation where baboons, leopard, and vultures live.[53] Many assume the Serengeti's grass and scattered trees as the typical savannah of early human experience, but in reality, savannah vegetation differs (and differed) enormously in its detailed composition.[54] It was never solely the home of the scattered Acacia thorn tree given by Orians and Heerwagen in their founding interpretation of savannah.

Savannah is so loosely described that it is not recognised in the UNESCO classifications of African vegetation.[55] The very term savannah covers a wide range, with woody cover (forest or woods) varying between 5 and 80 per cent, a massive variation in spatial type.

Teasing out exactly what a savannah might be is important in the light of the presumption of a certain type likened to the public park of green lawns and groups of scattered trees. Public parks might look like more wooded forms of 'savannah' as seen in the image below in Soweto (Figure 3.5), where the spaced trees are quite like the savannah of the Ugalla area in Tanzania shown in Figures 3.3 and 3.4.

Important too is the composition of the grass in the grassland component of a savannah, because major grass species found in the

Figure 3.5 Park in Soweto, South Africa, with students returning home from school on a warm February day. The park, as is mostly the case in southern Africa, provides both shade and business opportunities for small traders.

Image: The author.

savannah are tall and do not 'afford distant views'.[56] The variety of grasses in African savannahs currently varies between thirty to sixty species depending on location, with six to ten dominant species.[57] Antelope grass, *Echinochloa pyramidalis*, grows up to 500cm (i.e. 5 metres), hood grasses (*Hyperharria* species) up to 300cm, spear grass (*Heteropogon contortus*) to 90cm, wild oat grass (*Monocymbium ceresiiforme*) to 120cm, elephant grass (*Pennisetum purpureum*) to 450cm (4.5 metres), and many other grasses between 30 to 120cm in height. Such grasses provide a good environment for predators to hide, particularly in the wet season when the grasses are tall in the tropical grasslands, and herds disperse. These grasslands are a far cry from the manicured grasses of public parks, Repton-esque landscapes of England, and early views of the American mid-West that Orians referred to as savannah-like environments.[58]

Human evolution has been complex. In discussing work on the environmental context for human evolution carried out in the first decade of this century, Mark Maslin and Beth Christensen,[59] who researched environmental change in Africa over the last 10 million years, noted that during times of human evolution there have been massive changes with climate variability, the disappearance and reappearance of lakes, and changes in grassland components. Palaeoecology has long questioned the idea that grassland-dominated savannahs were the predominant environment of early hominin evolution.[60] The idea is now more strongly considered as not clear at all[61] or simply incorrect.[62] Both archaic humans and early *Homo sapiens* moved continually between landscape types. As they inhabited the ecotones between vegetation spatial types, moving between woodland and forest, and woodland and grassland, they would have had daily pressure to evade predators. In a study of buffalo predation, it was clear that lions try to stalk their prey from cover and prefer the ecotones between grasslands and woodlands.[63] The forest was not the area of threat as Orians presumed in 1980, but the spaces in between – the more open habitats of the grasslands. We might also consider the acacias of the grass and scattered trees; most have thorns and would have been savage for humans or early hominins to climb to escape predation (Figure 3.6).

Who are 'we' who evolved into 'us'?

For many years we believed that our abilities in construction were unique. In 1960 the Leakey family in Tanzania discovered *Homo habilis* ('the handy man') and the idea grew that *Homo habilis* displayed the beginning of stone toolmaking, when early members of the *Homo* genus emerged about 2.6 million years ago. This idea is now known

Figure 3.6 All African species of Acacia have thorns (hooks), or spines. Here the spines are white, but they can be dark as well; some species have recurved thorns (e.g. *Acacia mellifera*, or 'wait-a-bit thorn'); *Acacia nigrescens*, or nob-thorn, has prickles on its bark and no spines. Spines, recurved thorns, and prickly bark would have been a hurdle to tree-climbing for humans seeking shelter on the grassland-woodland savannah type.

Image: The author.

to be incorrect.[64] A more complex view is emerging. Many early hominins made tools. Toolmaking is of greater antiquity and dates to about 3.4 million years ago – far longer than we once supposed. In other words, toolmaking was not confined to us, or even hominins related to *Homo sapiens*. This shift in thinking is a major new understanding within anthropology that has altered its own theories of human behaviour in response.

In the 1960s, when the Leakey family made their discoveries in the Olduvai Gorge that captured the global imagination, the savannah landscape of that region was widely presumed to be the type of landscape in which these early hominins walked and on which human evolution occurred, linking human evolution with the dry Serengeti Plains. If we place in context these anthropological discoveries and ideas of the 1960s, they occurred before the revolution in the earth sciences that established plate tectonics and continental drift from the 1970s.[65] At that time the molecular age of DNA analysis was in its infancy, and new understandings of how climate has changed had not yet emerged.[66] All of these revolutions in science are of immense importance in telling the story of human evolution and the environmental history of landscapes. In the 1980s and 1990s, many new discoveries of hominin fossils yielded intense and radical changes in knowledge of hominin evolution and their environments. It is important to note the historic framing of savannah theory in landscape studies because landscape perception failed to incorporate these newer ideas into its own theoretical ideas. We need to ask – why not?

The hominin fossil record is now seen as complex and non-linear, with eleven new species and four new genera named since 1987.[67] Palaeoclimatologists and hominin experts make two important points about the explosion of new knowledge that has been gained from three major arenas of inquiry: new fossil discoveries, molecular evidence for links between species and ourselves and knowledge of our own more complex genome, and greater understanding of the climate of Africa. First, the hominin phenotype has a much greater range than previously thought, suggesting that early *Homo* and their ancestors were far more variable than previously supposed. Second, new methodologies in extensive and growing studies in molecular genetics and taphonomy (the study of how organisms are fossilised) have allowed a far more accurate linking of human phenotypes to the environments where they were living.[68] This is fabulous news for landscape architects as we are beginning to understand in more detail each decade the human story in our landscapes.

Our ancestors 'came down from the trees' approximately 1.6 million years ago (Ma) in what was probably a long and protracted change, beginning 1.8–2.1 Ma, long before *Homo sapiens* evolved. The move into many and various savannah forms of landscape occurred so long ago that it involved many archaic human types, including *Homo ergaster* (1.9–1.4 Ma). Archaic forms came down from the trees long before *Homo sapiens* evolved 200,000 years ago. The famous Lucy who was a member of *Australopithecus* of 3.2 Ma years ago was arboreal as well as bipedal. As Mary Leakey, who studied the original finds in the Olduvai Gorge noted: 'Our past is like that of any other mammal. [We have] a very complicated, diverse past with lots of different species; many of which became extinct.'[69]

About 1.6 million years ago *Homo erectus* moved out of Africa into the Middle East and onwards to Europe and eastern Asia. Many archaic species were ancestors of *Homo sapiens*, *Homo neanderthalensis*, *Homo erectus*, *Homo floriensis*, and the people known as 'Denisovans' of eastern Asia. The last common ancestor we share with *Homo neanderthalensis*[70] is *Homo heidelbergensis*, who was hunting large animals such as deer by at least 780,000 years ago in Europe and used hafted spear technology[71] 500,000 years ago.[72] Other human types living with *Homo sapiens* 60,000 years ago were *Homo erectus*, *Homo neanderthalensis*, the Denisovans, *Homo floriensis*, and an unknown hominin likely to have been in southern Asia and only recently noted due to aberrations in the DNA of Denisovans.

The many new ideas of human evolution, our interbreeding with other human types, and the increasing understanding of the cultural capacities of other human types, whether concurrent with us or prior to our evolution, provoke the question, by their very diversity, of what landscapes they walked on? Who walked on savannah?

What landscapes did we live in during archaic and prehistoric times?

All archaic and recent *Homo* species walked on various landscapes, ecotones lying between landscape types and mosaics, and lived in woodland and forest. At Schöningen, Germany,[73] on the edge of what is now a major lignite mine, an early *Homo* lived 300,000 years ago by a lake with a boreal, cool-temperate climate, in a 'mix of meadows and forest steppes with an abundance of pine, spruce and birch woodland'.[74] They lived with straight-tusked elephants, rhinoceros, red deer, bear, horse, water-voles, and an extinct beaver-like animal. The area yielded fine provisions of acorns, hazelnuts, raspberries, and reeds and sedges. This is not a savannah of any type but an environment with particular difficulties and opportunities that were clearly met with great ingenuity because this human type stayed here for about 200,000 years.

The type of landscapes on which various humans evolved, and their importance to our understanding of human history and evolution, remain points of vigorous debate in palaeoanthropology. For example, an eastern African study to determine the extent of woody cover in hominin environments over the last 6 million years showed that tropical savannahs have been present in eastern Africa for at least 6 million years, and that within the range of savannah types, open grassland was far less abundant than wooded grassland.[75] This work suggested that the variation in woody cover might have been important for behavioural adaptations. However, the work also suggests that open environments were typical of sites where our ancestor *Ardipithecus ramidus* lived – that is, more in keeping with a 'savannah hypothesis' of anthropology.

Other palaeoanthropologists have very much criticised this work for its use of soil data, which are considered the least able to delimit biological niches, and the interpretation of open environments.[76] They point out that the selection of soils in the study was entirely composed of poor soils that support only grasslands and not the richer soils that support woody trees. Critics consider this a skewed result that greatly underestimates the percentage of woody cover in all sites examined.[77] This problem of technique and interpretation is important for students of landscape (in its widest sense) to consider because it tells us that we need a great deal of assistance in our interpretation before adding a new theory, without criticism, into our repertoire. We need to keep up with the debates or be advised by experts.

The work described above considered the landscapes of *Ardipithecus ramidus*. It is important that we are clear as to which early hominin is being considered when perception researchers talk of 'our' early perceptions? – Is it the early *Austropithecus*, *A. ramidus*, *Homo erectus*, *Homo heidelbergensis*, the other *Homo* species (Neanderthals, Denisovans) who are now represented in our own genome, or just us, *Homo sapiens*, who evolved out of Mary Leakey's 'complicated, diverse past'?

The prevailing message from anthropology today is that our early, or archaic, hominins such as *Ardipithecus* lived in landscapes of forest margins to woodland, 'a niche that involved terrestrial bipedality and routine dependence on arboreal substrates and resources'.[78] These early hominins lived in mosaic habitats of forest patches and woodlands that 'graded into more open grasslands'.[79] A varied landscape mosaic like this is typical of a tropical rift valley, where 'tectonically generated topography creates altitudinal biotic lines, with vegetation patterned by temperature, rainfall, slope, drainage, soil texture, fertility, and (especially) water availability'.[80] Many mosaic and heterogeneous landscapes were within a day-range for individuals of our ancestors.[81]

Palaeoanthropologists have asked which landscapes these early hominins might have preferred. We cannot really answer that except to look to 'savannah chimpanzees', who express preferences for specific habitats and foods. Zoologists can interpret some idea from them despite the divergence of chimpanzees' ancestors and our own ancestors approximately 6 million years ago. While some zoologists criticise using chimpanzees as models for human behaviour and landscape selections,[82] these models can illuminate some aspects of life for early hominids.

So, how do chimpanzees behave? The chimpanzees at Ugalla in western Tanzania inhabit landscapes similar to those our ancestors inhabited long ago.[83] Today, chimpanzee parties move freely across savannah types ranging from open woodland to heavily forested.[84]

The authors of the savannah theory supposed that the savannah gave good shelter, hunting, and disease-free environments. Nest-building suggests otherwise. Early hominin nest-building plays a part in my inquiry wondering where we slept for safety. Whether in forest or savannah, great apes build nests, usually building a new nest every night.[85] Nest-building in great apes may be the foundation of constructivity in hominids.[86] Nest-building, common to both our ancestral apes and early *Homo* antecedents, is a trait that has probably existed for millions of years.[87] To establish a sense of sleep quality and the advantages of sleeping above the ground, the zoologist Fiona Stewart slept in trees with Senegal's Fongoli chimpanzee community.[88] She found (perhaps not surprisingly) greater sleep disturbance on the ground, due to concerns arising from vocalisations of terrestrial predators in the night, and a profound reduction in bites from mosquitoes, tsetse flies, and ants when aloft (Figure 3.7). Trees were important to survival because all *Homo* were hunted species and suffered about the same levels of predation as modern primates. They found safety from leopards and other cats in trees and dense underbrush, with daytime rests in dense vegetation.[89] With the invention of fire as early as 800,000 years ago, members of *Homo* might well have changed tree-sleeping to ground-sleeping.[90] Additionally, with varied environments, trees might not have been an option.[91]

In what environments did we evolve then? The consensus is that our archaic ancestor *Australopithecus* was eurytopic,[92] which means that they lived in diverse habitats. Matrix environments provided a wide diet

Figure 3.7 Zoologist Fiona Stewart of the University of Cambridge in a chimpanzee nest for the night. She found that she had a better night's sleep aloft than on the ground. All great apes sleep aloft.

Image: With permission, Fiona Stewart.

and opportunities. This eurytopic living is associated with a growing understanding that 'living on the edge' was important for success.[93]

New ideas in palaeoecology and anthropology have de-emphasised static reconstructions of early hominin habitats.[94] Palaeoanthropologist Richard Potts, director of the Smithsonian's Human Origins Program, stresses the need to examine the entire range of variability in shaping human evolution and to take into consideration all types of environmental fluctuation.[95] Researching changes to the African climate over the last few millennia, including the more recent past, has revealed that East Africa experienced rapid shifts from wet to dry conditions and back again. According to Richard Potts, 'important changes in stone technology, sociality, and other aspects of hominin behavior can now be understood as adaptive responses to heightened habitat instability'.[96] The key change in thought about human evolution is that climate variability may have provided a catalyst for evolutionary change,[97] not climate or landscape stability. This is a major point in our reconsideration of landscape perception. Climate variability meant that we were moving between types of landscapes over generations, and dealing with change, complexity, and variations in seasonality.

Single sites have fluxed and changed over time. Potts points out that there have been many attempts to pin down the one environmental type that was the key to human evolution.[98] But habitat-specific scenarios linked to evolution do not exist. Rather, habitats were unstable and subject to 'large-scale remodelling' over time, and Potts considers that 'enhanced cognitive and social functioning all may reflect adaptations to environmental novelty and highly varying selective contexts'.[99] Evidence from the fields of palaeobotany now shows that 'highly variable ecosystems accompanied the emergence and dispersal of the genus *Homo*'.[100] Parties of humans stayed in the same landscape as the landscape changed around them over one individual's lifetime or in that party's lifetime. They might have adapted to changing conditions by moving out of a particular site, or might have stayed where they were and adapted to new site conditions. In this way, our ancestors experienced a wide range of environments. It is instability, not stability, that led to the emergence of key changes of both human evolution[101] and human culture,[102] and this point is essential for us to ponder because we live in an age of climate change. As discussed by Mike Hulme in *Why We Disagree about Climate Change*,[103] pressure from shifting climate and landscape conditions has previously been a tremendous stimulus for innovation and societal adaptation and can 'accelerate the development of new complex civil and social structures'.[104] The idea that pressure produces innovation is a view shared by the great British historian Arnold Toynbee, who stated that 'ease is inimical to civilisation'.[105]

I suggest that savannah as a landscape type was no more important than others. Of far greater importance was our ancestors' need to comprehend new or changing landscapes and to engage with those conditions in different ways. What we learnt during our early history was not about how desirable a savannah landscape of scattered trees and grass was, but that we could survive and prosper in fluxing places with variable difficulties and benefits. If there is a message left to us from our history, it is not of a preferred vision for one specific type of landscape, but that we could enter new environments and cope, using known skills such as nest-building in trees, and no doubt learning new skills as we began to camp in new places, whether on the ground in early archaic times or on other types of sites. We could survive, manipulate, construct our environments, and meet our needs. Such daring to deal with difficulty has led us to accommodate our cultures to many types of landscapes all over the world, from the Arctic's risky cold to the intensely high-risk waterless desert regions. More recently, we risk the construction of systems that are returning land to previous states, such as the re-poldering of many sites in the Netherlands and the retreat from parts of New York City as a rethinking following Hurricane Sandy.

While I have emphasised the early hominins and early *Homo sapiens* in this discussion, it is opportune to consider the many different environments we have lived in more recently and where we remain. Ideas as to what these landscapes were like historically are also changing. In Europe, probably the best studied region since the last Ice Age, discussion about landscape history mimic debates concerning the far earlier environments of the Palaeolithic. In Europe, arguments centred on the idea that lowland Europe was once dominated by high, closed-canopy forest prior to major human impacts about 6,000BP.[106] A theory proposing a mix of wood and pasture countered this idea. The 'wood-pasture hypothesis' might not seem relevant to the current discussion of 'the savannah theory' but it was so strong that it led to management decisions and policies emphasising parklands, wood pasture, ancient trees and views[107] – all sounding quite like the visual expectations of the savannah theory of human perception – and these policies had impacts on designs for public open spaces. But recently, the tide seems in favour of the closed forest as the major biome of lowland Europe, largely from the evidence of pollen analysis. It appears that a forest with an open canopy only developed in the last 3,000 years.[108] The wood pasture theory has been abandoned,[109] and those previous management decisions now appear poor.

What impacts did new landscapes have on our minds?

The savannah theory suggests that our long descent from the trees left us with a preference for one particular landscape, which is a remarkable claim in view of the many landscapes we happily inhabit now. The savannah theory implies very strongly that we as a species have a closed cognitive experience despite the very many varied environments to which we have been exposed. Many questions arise as to what changes the acts of entering new environments created for us. What differences did we meet over time? Is the type of landscape per se important, or was the thinking encouraged by changes to landscape more important? How did we cope with change and resource heterogeneity? How did eurytopic environments impact our social structures? Were new dependencies on each other created, leading to different social structures and the need for alternative approaches to problems?

New ideas in cognitive anthropology suggest that, if the premise of the savannah theory were valid, we might not be capable of learning from different experiences; surely common sense tells us that this is not the case. An important new theory that I draw upon here is that of Material Engagement, as articulated by Lambros Malafouris in his beautiful book *How Things Shape the Mind*.[110] Written from the viewpoint of cognitive archaeology, he explores how we interact with the world and with material 'things' in the world, including objects that we make ourselves. Material things include our constructed landscapes, whether vast agricultural landscapes or ecosystems of the city. In Material Engagement, our thoughts, actions, and material things are inseparable. That is, thinking is a process of interactions of brain and body with the world.[111] In short, our landscapes have shaped our mind.

That landscapes can impact our cognitive experiences is related to new ideas in science concerning the very way in which genetic information is expressed in the body, and what controls that expression. For the last 150 years, humans have accepted that our genetic information is locked into our genetic code. Yet fifty years before Darwin published *On the Origin of Species*, the great French naturalist Jean-Baptiste Lamarck had hypothesised that one generation could pass down information to the next by a process referred to as 'soft inheritance' or use/disuse theory, or simply Lamarckism. Importantly, soft inheritance postulated that our environment could inform expression of information in human evolution. For most of the twentieth century, the scientific community rejected this concept as without foundation, and the idea was discredited. But discoveries this century have renewed the debate about the impacts of the environment on our genes. Fundamentally, the idea of 'soft inheritance' or 'inherited variation'[112] is that the environment can determine the functioning and expression of genes. This idea will be of great interest to designers because it suggests that the environment might be a far more profound arbiter of who we become, with subtler overarching impacts than imagined or articulated previously. Such an idea would feed into concepts of how we view and respond to environments, and how we might lose experiential feelings, or lose the sense of nature such as described by environmental generational amnesia,[113] and how we might gain responses different to those of our ancestors due to recent and personal experiences. Grandmothers will be pleased to know that in human evolution, the increasing longevity of humans and the subsequent presence of grandmothers in the social environment was a key to our ability to deal with change and stress, and increased the resilience of tribes.[114]

In landscape design, the links between landscape and brain are important because 'humans alone shape and reshape the environments that shape their brains'.[115] This is cultural evolution.[116] Bruce Wexler has written that cultural evolution through reshaping our environments has created 'incremental and widespread population variability ... stored in all aspects of cultural artifacts and practice'.[117] From the gathering information we are now acknowledging a human past characterised by a plastic climate, a plastic environment, plastic genes, and plastic cultural responses to changing landscapes.

Our brain is not what it used to be

Have we the same brain as early hominins or even earlier millennia of *Homo sapiens*? The core of this question is: has our brain changed? Recent work on the plasticity of the human brain confirms that the brain changes. Neurological research over the last twenty years has shown that individual brains have the capacity to 'relearn' wiring after some injuries, a phenomena termed neuroplasticity. Importantly in our story about our relationships to landscape, neurological science has also found that our brains are not the same as they once were. For the last 100,000 years, our brains have been constantly changing, and we do not have the same brains as those long-gone archaic ancestors who descended from trees and eventually came to walk on all the world's landscapes. Nor did we arrive at some settled modern 'endpoint' at any stage, and are now stopped, fully modern. We are still changing. That the brain changes is one of the most important findings in neurology in the last few decades.

Changes have occurred in the human psyche, and human genes alter due to inferred cultural selection pressures.[118] The apparent shaping of the human genome by culture is very important in our story, as it suggests that even the genome is somewhat plastic and *can be* shaped.[119] Many genes have responded very rapidly to cultural selection pressures; these include genes regulating milk digestion (lactose absorption following the domestication of cattle); cytochrome absorption for detoxification of newly domesticated plants; heat-shock and cold tolerance due to exposure to novel climates with human migrations and dispersals across the planet; changes in immunity through exposure to new environments and new pathogens; skin pigmentation, hair colour, and freckles have been local adaptations to levels of solar energy; brain function and the development of language skills and vocal learning as we evolved our social intelligence; and skeletal development. Even tooth-enamel thickness has changed due to the physiological pressures from the invention of cooking. There are two striking things here: first, that a lot has happened in the last 10,000–15,000 years, and second, that these changes in the human genome show that we have changed physiologically and genetically in response to new landscapes and new situations to which we have been exposed. This plastic and responsive process has not ended.

The newness of genes for responses like milk drinking[120] has changed the way we think about genes. We have 21,000 essential, protein-coding genes. To put this in context and make us modest, this is less than half that of rice, and only a tiny bit more than the house mouse. This 21,000 is only about 1.5 per cent of our total genome;[121] that is, only 1.5 per cent is expressed[122] of 2.9 billion base-pairs.[123] Until very recently we thought that the mass of the genome was 'noise' and of no use to us at all, or that it represented discarded material lost through evolution. Geneticists now believe that the apparently unused part of our genome can come into play under new conditions and stress, and we can change quite rapidly in response to changing landscapes.[124]

Similarly, neurologists believe that our brain can adapt to changing social, cultural, and environmental conditions. Just as work on the human genome has been a revolution in anthropology and medicine, the metaplasticity of the brain has been a revolution in neurological science and has in turn impacted the way in which anthropologists view human evolution. The cognitive anthropologist Lambros Malafouris notes:

> If the intrinsically plastic human brain undergoes constant change subject to various developmental, environmental, and cultural factors, it cannot simply be assumed that 'anatomically modern human intelligence' refers to a fixed and stable speciation event ... the hallmark of human cognitive evolution is metaplasticity – that is, ever-increasing extra-neural projective flexibility that allows for environmentally and culturally derived changes in the structure and functional architecture of our brain.[125]

This notion is clearly a long way from a set brain that has not changed since our ancestors walked on savannah.

New ideas of how each generation reframes its environments underline the movement of human thinking and preferences for landscape types. Bruce Wexler[126] wrote of the ways in which our brain works in relation to culture and noted that ideas pertaining to brain development from birth have changed remarkably in the last few decades:

> Once internal structures are established, they turn the relationship between internal and external around. Instead of the internal structures being shaped by the environment, the individual now acts to preserve established structures in the face of environmental challenges, and finds changes in structure difficult and painful ... Individuals seek out stimulation that is consistent with their established internal structures, and ignore, forget, or attempt to actively discredit information that is inconsistent with these structures.[127]

This passage suggests that local, personal experience impacts us enormously and is passed on from grandparents to parents to children. Some studies on landscape preferences have clearly shown that what we know of when young leaves a lasting legacy. A Swedish study found that people feel more at home with the type of landscape they grew up in and prefer qualities connected to those early landscapes.[128] We need to note, of course, that each new generation experiences new things and adapts to them, and this is why each generation experiences the world with slightly different eyes.[129] Moreover, the process of generational change is likely to have accelerated in the last few hundred years. It is probable that in the prehistoric world change occurred more slowly unless due to catastrophic events.

It is now clear that the brain learns from experience and is plastic. The capability of acquiring new knowledge is different for every person due to that person's history and experience. Neuroscience is now revealing finer differences.[130] The move into any new landscapes – whatever they were and whenever they have occurred, whether 30,000 years ago when the first Americans entered those continents or the migrant experiences of the living African migrating to Europe now – gave us and continues to give us new ways of thinking as we adapted to our new situations and learnt from them, passing on knowledge, and likely leaving familiar preferences for each generation, even though that 'familiar' changed generation by generation.[131]

What and how were we thinking?

Since we have no way of knowing the hominin mind, we cannot make any assumptions about the impact on cognition for preferences, such as a savannah preference, or what the hominins were thinking. Instead, we can ask the question of *how, rather than what*, the hominin was thinking.[132] Were early hominins thinking the same way as we do now?

If we consider the savannah theory, it assumes that early Homo developed a preference for a certain type of landscape – an opinion about what they liked. But, if our brain has changed, it is important to also ask if we might think differently about individual and collective subjectivity – that is, about 'we, us, me, and you', and any opinions within those groups. As the archaeologist Colin Renfrew noted, we cannot assume the same way of thinking;[133] what seems normal now would have been alien or odd previously. Renfrew made the comment in relation to living in cities, which now seems so normal to at least half of humanity. Yet living in cities is very recent. Forgetting that ideas change can 'blind us to the uniqueness of our own sense of self'[134] as it is today. We have no testimonies from the unwritten past. Therefore, we cannot know what our unlettered ancestors were really thinking.

When we have tried to imagine past thinking, we can be incorrect. When paintings of leopard-spot horses, one of the staple foods of Palaeolithic humans,[135] were found in a cave in France, some archaeologists imagined that the spots represented astronomical or spiritual associations with horses. However, leopard-spotted horses were one of four colour-types found in the Palaeolithic,[136] and the cave art shows the horses as they lived. The existence of contemporary 'leopard-spotted' horses suggests that the spots painted in the cave were not an astronomical representation but a true depiction of the real spotted horses of ancient times. The image of Figure 3.8 shows the Appaloosa breed, with their leopard spots.

New ideas in Greek philology suggest that we need to consider the evolution and changes of cognitive thought – that how we thought about ourselves, or more precisely our *self*, might have changed, even over the few thousands of years since Archaic Greece. Changes in how we think about ourselves suggest that there were surely changes in our *personal* responses to landscapes, and if such personal responses as we now understand them, even existed. Can we know what people thought about themselves long ago? Again, any shifts in our thinking of self, ego (ἐγώ), personal space, and the interior life, are largely lost to us if we have no written material to examine.

Linguists have begun to look at the evolving use of grammatical objects for self-reflection in old languages such as Archaic Greek,[137] which sits within the group of languages known as Proto-Indo-European. Such an

Figure 3.8 The Panel of the Dotted Horses as seen in the Pech Merle Caves in France dated to approximately 18,500–25,000BP (top) and Appaloosa horses today in New Zealand (lower image).

Sources: Upper image with permission from Patrick Cabrol, Centre de Préhistoire du Pech Merle; akg-images, and lower image with permission from Jeanine McDonald, Mararoa Appaloosas, Invercargill, New Zealand.

examination is surely a part of this story of how we think about ourselves in relation to landscapes, whether it was as a *self*, *me*, *mine*, or as a group. This story might be strongly related to asking if long ago we thought 'What is good for *me*?' or 'What do *I* prefer?' in addition to 'What is good for us as a group?' Or we might not have had any thoughts of 'self'. In examination of Proto-Indo-European, the linguist and philosopher Edward Jeremiah found that the idea of *me*, *myself*, *self*, and *psyche* is quite recent, at least in Ancient Greece. He determined this from his analysis of the Archaic Greek use of reflexive – i.e. me-centred dialogues – word usage. Jeremiah noted: 'The Greek relationship with reflexivity will come to condition the evolution of the West, and especially modernity, for which the definition of the individual as a reflexive agent, a self, becomes a guiding cynosure.'[138] It is important to consider that our current sense of self is unlikely to have been the normal human condition, and personal relationships to landscape types, or personal preferences for landscape types, are likely to be a relatively modern idea grown out of this guide. How might 'we' have considered landscape long ago then? We do not know, yet the savannah theory assumes that we should be something like the mentality of the Australophicines, *Ardipithecus*, or the early archaic *Homo* who came down from the trees. However, we are not.

The savannah theory also assumes our common response to landscapes. Recent work that analysed approaches at the macro level has suggested that our cultural capacity increased between 200,000 to 30,000 years ago, due to increasing behavioural flexibility. Humans of this period 'were not specific in their habitat choice', and flexibility in behaviour was the single most successful adaptation, including the 'flexibility with which people accessed different parts of their landscape'.[139] These responses to landscape were not uniform and were not accumulative. Behavioural flexibility in our ancestors was the most important human adaptation to landscape. Flexibility applied at both the local level, with tribes using matrix environments, and at the population level, with different populations inhabiting quite variable climatic regions.

Perhaps the closest idea we can get to how humans might have *thought about* landscape comes from the indigenous people of Australia who have maintained ancient links with their landscapes. Indigenous feelings for Country do not place landscape as an object to be viewed. Rather, Country is a subject, not an object. Each particular member of the tribal group has a totem, a part of that living landscape for which they are custodians of stories and their contained knowledge, and in this way the self is also inseparable from Country.[140] For example, the local Anangu, who are custodians of Uluru in central Australia, do not want tourists to climb 'the Rock' or to take photos of its changing colours at sunset, but do want people to walk in the landscape as a spiritual experience. In contrast, many landscape researchers objectify places by asking for landscape responses to an image taken with a lens and formed into a two-dimensional photo. Essentially, it is Country as object that we see in many studies of landscape perception, and these studies are also based on me/myself that objectifies landscape to the point where a three-dimensional living and spiritual place is diminished in a photo. The two-dimensional masquerades as landscape, which it can never be, and the photo represents an extreme form of narcissism because it positions the viewer outside, and aloof from, all landscapes and all creatures and plants that live within, even other people. It is likely that this photographic view of landscape as object is akin to nothing in past human history.

Conclusions

Evidence I have presented here is that human nature is flexible, ongoing, becoming, and was never static at one point of experience or learning. Human learning and response have been a continual dynamic process and a key feature of human evolution, and reveal the 'profound plasticity of the human mind'.[141] The savannah theory positions itself outside processes and sees a particular period in time as having permanent impact. In short, the move into the savannah has been seen as a static moment in time that set our psychological thinking about landscapes and our physical responses to them to this day. The idea of stasis appears to be a result of the pressure inherent in Western philosophy for the last few hundred years of ignoring process, change, and becoming in favour of uniformity, being, and permanence.[142]

A powerful picture emerges from this inquiry. Human preference is more complex and culturally attuned than landscape perception studies have assumed. Human evolution was a complex movement over time and space between various landscapes, including those that have and had neither forest nor savannah. We inhabited many landscapes of scale, ecotones, mosaics, and seasons, and we continue to do so; Neanderthals also lived in dissected terrains, giving them strong contact with various ecotones, landscape types, and food sources.[143] The lived experience of our ancestors exposed them to many environments from which we have learnt the capacity to change as fundamental to human being; variation, flux, change, and adaptation are the main and most important 'imprints' we received. Each new landscape we entered was all novel to some degree and required thinking and responses to new problems. Anthropologists have been noting the impacts of our changing environments on human technologies and are now asking 'to what extent is technology dictated, or even selected, by environmental fluctuations?'[144] I propose a new concept of *shifting adaptabilities* – responses to changes in climate and food opportunities and other difficulties that have enabled humans to occupy almost every biome on Earth and respond culturally in many ways.

How did we adapt to change? What does the clear success of our adaptation tell us about our capacities and, more importantly, how can we learn from this past to deal with coming change in the next decades and centuries? *Shifting adaptabilities* suggests a more productive and hopeful view than the static view of human thought. The challenge at this time in history is that we now have 9 billion people, and dealing with our population is the central challenge.

The enduring myth of the savannah theory has dominated the literature about our relationship to landscapes. It has fostered an unreflective intellectual climate, and may well have become so entrenched that experimentation on human preferences for ecologies and views in our constructed world no longer really tests the null hypothesis that the savannah type of grassland and scattered trees was of no more importance than any other landscape that could afford us some succour.[145] New information suggests that, if we learnt anything from living on savannahs and all other landscapes, we learnt that we could deal with changing environments and instability. Change really is the one constant in human history.

Shifting resiliencies and adaptabilities are the key to landscape perception and use. The idea of a 'savannah theory,' with the savannah as the main driver of human perception, must now be cast aside because of scholarship from anthropology, cognitive archaeology, neurology, palaeobotany, and genetics. Just as new knowledge in human genetics has begun to generate new hypotheses about the history and pattern of biological variation within human populations,[146] we need to also revise the impacts of landscapes upon human histories. The human experience through history has been broad, with *shifting adaptabilities* and shifting perceptions because the mind is plastic. I suggest that the overriding features that have underlain our relationships to landscapes are plasticity, fluidity, and our capacity to adapt to new situations and to prosper. Indeed, that the emergence of key adaptations in human development occurred when we were placed under high climate variability suggests that the pressure to change again will place evolutionary challenges upon us. How will we as constructors of this world respond?

Notes

1 Friedrich Nietzsche (1881), 1999 and 2011, *Morgenröte*, ed. Giorgio Colli and Mazzino Montinari, Berlin: Verlag de Gruyter, aphorism 168, lines 21–22.

2 The earliest Neanderthal remains found in Gibraltar in 1848 had been considered to be of deformed humans. Only after the finds in Germany did a reassessment of those bones reveal the same type of human.
3 Gordon Orians and Judith Heerwagen, 1992, 'Evolved responses to landscapes', in Jerome H. Barkow, Leda Cosmides, and John Tooby (eds), *Adapted Mind: Evolutionary Psychology and the Generation of Culture*, Oxford: Oxford University Press, pp. 555–598.
4 Ibid.
5 Ibid.
6 Ibid.
7 Eugene E. Harris, 2015, *Ancestors in Our Genome: The New Science of Human Evolution*, Oxford: Oxford University Press.
8 Genes from Neanderthals contribute about 1–4 per cent 'of the ancestry of people outside sub-Saharan Africa, and Denisovans contribute 1–6% of the ancestry of people in island South-east Asia and Oceania'. As discussed in John Hawks, 2013, 'Significance of Neandertal and Denisovan genomes in human evolution', *Annual Review of Anthropology* 42: 433–449.
9 I make this point because meetings between *Homo sapiens* and other human types would not have been as great a cultural shift as, for example, the meeting of English gentlemen soldiers and indigenous cultures in the Americas, Africa, and Australia, and perhaps less dangerous. In 1977, a group of nomad Aboriginals of the Mandildjara people of the Gibson Desert who had had no contact with European settlers met 'Europeanised' Aboriginals who had clothes and cars; the nomad young men (who had no women) were deeply interested in the young women as women, not that they were different, and the young women were deeply impressed with such fit, virile young men. This meeting is recorded in W.J. Peasley, 2009, *The Last of the Nomads*, Fremantle: Fremantle Arts Centre Press. How might a single, 'wife-less' *Homo sapiens* male view a lost female Neanderthal woman? How might a single female have viewed a virile male? We will never really know, but part of the answer is in our genes. Companionship might well have been a strong driver.
10 Sriram Sankararaman, Nick Patterson, Heng Li, Svante Pääbo, and David Reich, 2012, 'The date of interbreeding between Neanderthals and modern humans', *PLOS Genetics* 8: e1002947.
11 Kay Prüfer, Fernando Racimo, Nick Patterson, Flora Jay, Sriram Sankararaman, *et al*. (many), 2014, 'The complete genome sequence of a Neanderthal from the Altai Mountains', *Nature* 505: 43–49.
12 Proof that we interbred successfully with other human types raises the very question of what a species is, and raises the question as to what ranges and variation a 'species' might have. However, livestock, or dogs and cats, shows the phenotypical variation possible in one species. It is an unanswered question as yet, although botany deals with the same issues in regard to hybridisation between species.
13 An example is the report on spears by *The Guardian* newspaper in the UK, 'Stone me! Spears show early human species was sharper than we thought', 15 November 2012. *Homo ergaster*, reported in this article, is a common ancestor of both *Homo sapiens* and Neanderthal.
14 See João Zilhão, 1998, 'Neanderthal acculturation in Western Europe? A critical review of the evidence and its interpretation', *Current Anthropology* 39: Special Issue The Neanderthal Problem and the Evolution of Human Behavior: S1–S44. In the responses to this paper, the placing of Neanderthals as 'others' can be clearly seen, and the presumption that they were very different to humans considered 'modern' (i.e. us).

15 The history of the discovery of the Neanderthals is given in Eugene Harris, 2015, *Ancestors in Our Genome: The New Science of Human Evolution*, Oxford: Oxford University Press.
16 This remains a source of intrigue, with some suggesting that the evidence is there and we are 'setting the bar too high' for something that might have to be inferred. See: Michael Balter, 2012, 'Did Neanderthals really bury their dead?', *Science* 337: 1443–1444; this outlines work done at La Ferrassie, in the Dordogne, France, where seven Neanderthals appear to have been buried together. Mortuary rituals are considered a deeply spiritual, human activity.
17 Indirect evidence from anatomy, archaeology, and DNA suggests some language abilities for Neanderthals, 'if not necessarily full modern syntactic language'. Sverker Johansson, 2015, 'Language abilities of Neanderthals', *Annual Review of Linguistics* 1: 311–332.
18 Davorka Radovčić, Ankica Oros Sršen, Jakov Radovčić, and David W. Frayer, 2015, 'Evidence for Neandertal jewelry: Modified white-tailed eagle claws at Krapina', *PLOS ONE* 10(3): e0119802. Available online at doi:10.1371/journal.pone.0119802.
19 Amanda G. Henry, Alison S. Brooks, and Dolores R. Piperno, 2014, 'Plant foods and the dietary ecology of Neanderthals and early modern humans', *Journal of Human Evolution* 69, 44–54. See also: Luca Fiorenza, Stafano Benazzi, Amanda G. Henry, Domingo C. Salazar-Garcia, Ruth Blasco, *et al.*, 2014, 'To meat or not to meat? New perspectives on Neanderthal ecology', *American Journal of Physical Anthropology* 156(S59): 43–71.
20 The emergence of 'behavioral modernity' was 'not a species-specific phenomenon' but the 'cognitive basis must have been present in the genus *Homo* before the evolutionary split between the Neandertal and modern human lineages'. João Zilhão, 2007, 'The emergence of ornaments and art: An archaeological perspective on the origins of "behavioral modernity"', *Journal of Archaeological Research* 15(1): 1–54. Further, cognitive development has not been linear, but is a diverse branching tree; see Marco Langbroek, 2014, 'Ice age mentalists: Debating neurological and behavioural perspectives on the Neandertal and modern mind', *Journal of Anthropological Sciences* 92: 285–289.
21 A.W.G. Pike, D.L. Hoffmann, M. García-diez, P.B. Pettitt, J. Alcolea, *et al.*, 2012, 'U-series dating of Paleolithic art at 11 caves in Spain', *Science* 336: 1409–1413. These caves contain paintings dated to 40.8 thousand years for a red disk, and 37.3 thousand years for a hand stencil.
22 Paola Villa and Wil Roebroeks, 2014, 'Neandertal demise: An archaeological analysis of the modern human superiority complex', *PLOS ONE* 9(4): e96424.
23 Richard E. Green, *et al.* (many authors), 2010, 'A draft sequence of the Neandertal genome', *Science* 328: 710–722; Kay Prüfer, *et al.* (many authors), 2014, 'The complete genome sequence of a Neanderthal from the Altai Mountains', *Nature* 505: 43–49.
24 A comment made by Daniel Adler, 2011, in response to John Shea's paper 'Homo sapiens is as Homo sapiens was', *Current Anthropology* 52: 1–35. Adler's comments on p. 16.
25 The term 'hominins' refers to *Homo sapiens* and all our fossil ancestors. This is distinct from hominids, a term that includes us, our fossil ancestors, and the great apes of today.
26 Mark A. Maslin, Chris M. Brierley, Alice M. Milner, Susanne Shultz, Martin H. Trauth, and Katy E. Wilson, 2014, 'East African climate pulses and human evolution', *Quaternary Science Reviews* 101: 1–17, outlines the basic

four main stages of human evolution as being: (i) proto-hominins such as *Ardipithecus* between 4 and 7 million years ago; (ii) *Australopithecus* about 4 million years ago; (iii) appearance of the genus *Homo* about 1.8–2.5 million years ago, and (iv) appearance of anatomically modern humans about 200,000 years ago.
27 Gordon H. Orians, 1980, *Some Adaptations of Marsh-Nesting Blackbirds*, Princeton, NJ: Princeton University Press.
28 Gordon H. Orians, 1980, 'Habitat selection: General theory and applications to human behavior', in J.S. Lockard (ed.), *The Evolution of Human Social Behavior*, New York: Elsevier, pp. 49–66; Gordon H. Orians and Judith H. Heerwagen, 1992, 'Evolved responses to landscapes', in J.H. Barkow, L. Cosmides, and J. Tooby (eds), *The Adapted Mind: Evolutionary Psychology and the Generation of Culture*, New York: Oxford University Press, pp. 559–579.
29 Like the early European settlers in Australia, the new arrivals in the 'New World' did not realise that this was in fact an old world and that these grassy plains had been largely constructed by the ambition and knowledge of local indigenous tribes to facilitate the management and capture of game. In the case of North America this was the bison, which had been managed for perhaps 10,000 years, and in Australia grassy plains were largely established on the better soils for kangaroo hunting. For a discussion on Australia's constructed hunting grounds see Bill Gammage, 2011, *The Biggest Estate on Earth: How Aborigines Made Australia*, Sydney: Allen & Unwin.
30 Marcel Hunziker, Matthias Buchecker, and Terry Hartig, 2007, make the distinction between theories and studies regarding landscape perceived as a space, such as the 'savannah theory' of Gordon Orians, and those perceived as 'place', in F. Kienast, O. Wildi and S. Ghosh (eds), *A Changing World: Challenges for Landscape Research*, Dordrecht: Springer, pp. 47–62
31 Even today, many human societies of African landscapes live in small communities surrounded by thorn hedges to protect both human and animal stock from predation.
32 G.H. Orians and J.H. Heerwagen, 1992, 'Evolved responses to landscapes', in J.H. Barkow, L. Cosmides, and J. Tooby (eds), *The Adapted Mind*, Oxford: Oxford University Press, pp. 555–579.
33 Although all the African *Acacia* were reclassified after the 17th International Botanical Congress in Vienna in 2005, this was extremely controversial, and although Australian *Acacia* kept the appellation due to the weight of species numbers – about 1,300 compared to Africa's 142 – this genera's name is surely too completely wrapped up in the culture and history of both continents for a name change based on genetics; like many, I believe that the name *Acacia* for both African and Australian species preserves the ancient link between the vegetation of these two very closely aligned continents.
34 See for example, chapters in Peter Howard, Ian Thompson, and Emma Waterton (eds), 2013, *The Routledge Companion to Landscape Studies* (Abingdon: Routledge): Catherine Ward Thompson, 'Landscape perception, environmental psychology', p. 29; Peter Howard, 'Perceptual lenses', p. 44; and Isis Brook, 'Aesthetic appreciation of landscape', pp. 111–112.
35 Richard Potts, 1998, 'Environmental hypotheses of hominin evolution', *Yearbook of Physical Anthropology* 41: 93–136. In this paper Potts outlines all the major extant theories regarding the drivers of human evolution and notes that many theories are driven by assumptions or little data. It is worth citing his findings that: 'Global environmental records for the late Cenozoic and specific records at hominin sites show the following: 1) early human habitats were subject to large-scale remodeling over time; 2) the evidence for

environmental instability does not support habitat-specific explanations of key adaptive changes; 3) the range of environmental change over time was more extensive and the tempo far more prolonged than allowed by the seasonality hypothesis; and 4) the variability selection hypothesis is strongly supported by the persistence of hominins through long sequences of environmental remodeling and the origin of important adaptations in periods of wide habitat diversity. Early bipedality, stone transport, diversification of artifact contexts, encephalization, and enhanced cognitive and social functioning all may reflect adaptations to environmental novelty and highly varying selective contexts.' These important findings appear to have not been followed by landscape architects.

36 Terry Hartig, *et al.*, 2011 make this point – of the need to pay attention to evolutionary theory and related research, but do not critique or offer an alternative view of the theories long-held. Terry Hartig, Agnes E. van den Berg, Caroline M. Hagerhall, Marek Tomalak, Nicole Bauer, *et al.*, 2011, 'Health benefits of nature experience: Psychological, social and cultural processes', in Kjell Nilsson, Marcus Sangster, Christos Gallis, Terry Hartig, Sjerpde Vries, *et al.* (eds), *Forests, Trees and Human Health*, New York: Springer, pp. 127–168.

37 In *The Two Cultures* (Cambridge University Press, 1949) and *The Two Cultures: A Second Look* (1964), C.P. Snow famously noted the divide between the arts and sciences.

38 Raymond A. Dart, 1925, '*Australopithecus africanus*: The man-ape of South Africa', *Nature* 115: 195–199.

39 See: Clifford J. Jolly, 1978, *Early Hominids of Africa*, London: Duckworth; and Richard G. Klein, 1989, *The Human Career: Human Biological and Cultural Origins*, Chicago: University of Chicago Press.

40 Discussed by Mark A. Maslin, *et al.*, 2014, 'East African climate pulses and early human evolution', *Quaternary Science Reviews* 101: 1–17, who note that the savannah theory was 'refined as the aridity hypothesis, which suggested that the long-term trend towards increased aridity and the expansion of the savannah was a major driver of hominin evolution'.

41 Renato Bender, Phillip V. Tobias, and Nicole Bender, 2012, 'The savannah hypotheses: Origin, reception and impact on palaeoanthropology', *History and Philosophy of the Life Sciences* 34: 147–184.

42 Tim D. White, Berhane Asfaw, Yonas Beyene, Yohannes Haile-Selassie, C. Owen Lovejoy, *et al.*, 2009, '*Ardipithecus ramidus* and the paleobiology of early hominids', *Science* 326: 75–86.

43 Richard Potts, 1998, 'Environmental hypotheses of human evolution', *Yearbook of Physical Anthropology* 107: 93–136.

44 Philip R. Nigst, Paul Haesaerts, Freddy Damblon, Christa Frank-Fellner, Carolina Mallol, *et al.*, 2014, 'Early modern human settlement of Europe north of the Alps occurred 43,500 years ago in a cold steppe-type environment', *Proceedings of the National Academy of Sciences* 111(40): 14394–14399.

45 Bryan Shorrocks, 2007, *The Biology of African Savannahs*, Oxford: Oxford University Press, p. 1.

46 As noted by Shorrocks, ibid., the term savannah or savanna is most likely derived from the sixteenth-century Spanish *zavanna* ('treeless plain'), and 'shares its origin with other well-known words such as barbeque, cannibal and papaya'.

47 Mahesh Sankaran, Nial P. Hanan, Robert J. Sholes, Jayashree Ratnam, David J. Augustine, *et al.*, 2005, 'Determinants of woody cover in African savannas', *Nature* 438: 846–849. Maximum tree cover is obtained at

approximately 650mm annual precipitation, and no trees below approximately 100mm per annum.
48 Frank White, 1983, *The Vegetation of Africa*, Paris: UNESCO.
49 Shorrocks, 2007, p. 23. Shorrocks notes that miombo has poor soils and poor nutrition for vegetation, and the predators are lion, leopard, cheetah, two species of hyena, wild dogs and jackals; primates are the chimpanzee in the forest savannah mosaic, and various species of monkeys across various types of savannah.
50 Jayashree Ratnam, William J. Bond, Rod J. Fensham, William A. Hoffmann, Sally Archibald, *et al.*, 2011, 'When is a "forest" a savanna, and why does it matter?', *Global Ecology and Biogeography* 20: 653–660.
51 Bryan Shorrocks and William Bates, 2015, *The Biology of African Savannahs*, 2nd edn, New York: Oxford University Press, p. 5.
52 The word Serengeti is derived from 'endless plain' in Maasai language.
53 Shorrocks and Bates, 2015.
54 Ibid., p. 5.
55 F. White, 1983, *The Vegetation of Africa*, Vol. 20, Paris: UNESCO.
56 Orians and Heerwagen, 1992, p. 558.
57 Shorrocks, 2007, pp. 40–42, discusses the grasses of the savannah and outlines the heights of the principal grass species in Africa.
58 Gordon H. Orians, 1980, 'Habitat selection: General theory and applications to human behavior', in J.S. Lockard (ed.), *The Evolution of Human Social Behavior*, New York: Elsevier, p. 60.
59 Mark A. Maslin and Beth Christensen, 2007, 'Tectonics, orbital forcing, global climate change, and human evolution in Africa: Introduction to the African paleoclimate special volume', *Journal of Human Evolution* 53: 443–464.
60 See E. Vrba, 1985, 'Environment and evolution: Alternative causes of the temporal distribution of evolutionary events', *South African Journal of Science* 81: 229–236.
61 Kaye E. Reed, 1997, 'Early hominid evolution and ecological change through the African Plio-Pleistocene', *Journal of Human Evolution* 32: 289–322. Kaye Reed examined extinct, early hominids; she found that '*Australopithecus* species existed in fairly wooded, well-watered regions', and '*Paranthropus* species lived in similar environs and also in more open regions, but always in habitats that include wetlands'. Also see: Lee-Thorp, *et al.*, 2007, 'Tracking changing environments using stable carbon isotopes in fossil tooth enamel: An example form the South African hominin sites', *Journal of Human Evolution* 53: 595–601.
62 Richard Potts, 1998, 'Environmental hypotheses of hominin evolution', *Yearbook of Physical Anthropology* 41: 93–136.
63 Herbert H.T. Prins (1996) studied buffalo predation by lions in *Ecology and Behaviour of the African Buffalo: Social Inequality and Decision-Making*, London: Chapman & Hall. The author discusses the use of the ecotone as the starting place for attack by lions on p. 136.
64 Stone tools discovered in Dikika, Ethiopia in 2010 are 3.4 million years old, and are therefore too ancient to have been shaped by *Homo* species; these are attributed to *Australopithecus afarensis*; see Shannon P. McPherron, Zeresenay Alemseged, Curtis W. Marean, Jonathan G. Wynn, Denné Reed, *et al.*, 2010, *Nature* 466: 857–860. This suggests earlier manufacturing of stone tools than previously ever considered. Further, several hominin species coexisted during this time period, as noted by Yohannes Haile-Selassie, Luis Gibert, Stephanie M. Melillo, Timothy M. Ryan, Mulugeta Alene, *et al.*, 2015, 'New species from Ethiopia further expands Middle Pliocene hominin diversity', *Nature* 521: 483–488.

65 Though Alfred Wegener (1880–1930) had suggested continental movement much earlier in his 1915 publication *The Origin of Continents and Oceans*. Wegener outlined evidence from several fields to suggest a new theory of why plants and geological patterns corresponded in continents separated by large oceans.
66 An early book of climate change and human history was Hubert H. Lamb's 1982 classic *Climate, History, and the Modern World*, London: Routledge; Hubert Lamb founded the Climate Research Centre at the University of East Anglia.
67 Mark A. Maslin, Susanne Shultz, and Martin H. Trauth, 2015, 'A synthesis of the theories and concepts of early human evolution', *Philosophical Transactions of the Royal Society B* 370: 20140064. Available online at http://dx.doi.org/10.1098/rstb.2014.0064.
68 Ibid.
69 Quoted in the Museum of the Cradle of Humanity, north of Johannesburg, South Africa, 2014.
70 *Homo neanderthalensis* and *Homo sapiens* are believed to have diverged somewhere between 800,000 to 400,000 years ago; Jayne Wilkins, Benjamin J. Schoville, Kyle S. Brown and Michael Chazan, 2012, 'Evidence for early hafted hunting technology', *Science* 338: 942–946.
71 The significance of hafted spears is that they are a technological advance on stone cutting tools or a simple wooden shaft with a pointed end. A hafted spear joins the two technologies of wood and stone, including the skill and science of binding the two, and suggests first intelligence of design and, second, that they organised and pre-thought out hunting, hinting at speech and good communication.
72 Jayne Wilkins, Benjamin J. Schoville, Kyle S. Brown, and Michael Chazan, 2012, 'Evidence for early hafted hunting technology', *Science* 338: 942–946. Hafting involves the dual technology of stone and wood and, if *Homo heidelbergensis* used them, it is considered that so did their descendants, both *Homo sapiens* and the Neanderthals.
73 The Schöningen site is known for the discovery of spruce and pine spears, dated now to 300,000 years ago, that could bring down a horse at 35 metres. Originally reported by Hartmut Thieme, 1997, 'Lower Palaeolithic hunting spears from Germany', *Nature* 385: 807–810. The 1997 ideas have been somewhat changed, as reported in *Science* 344, 6 June 2014, by Michael Balter in an article 'The Killing Ground'.
74 Thieme, 1997.
75 Thure E. Cerling, Jonathan G. Wynn, Samuel A. Andanje, Michael I. Bird, David Kimutai Korir, Naomi E. Levin, William Mace, Anthony N. Macharia, Jay Quade, and Christopher H. Remien, 2011, 'Woody cover and hominin environments in the past 6 million years', *Nature* 476: 51–56.
76 Tim D. White, 2014, 'Reply to Cerling, *et al.*', *Current Anthropology* 55: 471–472.
77 Gen Suwa and Stanley H. Ambrose, 2014, 'Reply to Cerling, *et al.*', *Current Anthropology* 55: 473–474.
78 Ibid., p. 473.
79 Ibid.
80 Tim White, 2014, p. 471.
81 Ibid.
82 One criticism is that the interpretation of chimpanzees as having societies close to ours and that they are taken as the only models of our close relatives. For example, Amy Parish, Frans B.M. De Waal and David Haig, 2000, comment that bonobos are equally close relatives to us and have very different social structures than chimpanzees. See: 'The other "closest living relative": How bonobos (*Pan paniscus*) challenge traditional assumptions

about females, dominance, intra- and intersexual interactions, and hominid evolution', *Annals of the New York Academy of Sciences* 907: 97–113.

83 A. Gibbons, 2009, 'Ardipithecus ramidus: Habitat for humanity', *Science* 326: 36–40. Of note in the *Science* article is that the existing landscape is very barren today but reconstruction shows a very different picture, of a woodland environment.

84 Hidexhi Ogawa, Midori Yoshikawa, and Gen'ichi Idani, 2013, 'The population and habitat preferences of chimpanzees in non-protected areas of Tanzania', *Pan African News*, accessed 28 May 2016. Chimpanzees in western Tanzania mainly live in miombo woodlands dominated by deciduous trees and some evergreen riverine forests. The authors noted that forests provide chimpanzees with safe sleeping sites; they have fewer grasses and more trees. See also: D. Moyer, *et al.* (9 others), 2006, *Surveys of Chimpanzees and Other Biodiversity in Western Tanzania*, accessed 28 May 2016.

85 Some apes, such as orang-utans in Borneo, use tree species with known anti-mosquito properties in the freshly broken-off branches. See C.J. Largo, 2009, 'Mosquito avoidance drives selection of nest tree species in Bornean orang-utans', *Folia Primatologia* 80: 163–163.

86 Fiona Stewart, 2011, 'The evolution of shelter: Ecology and ethology of chimpanzee nest building', PhD thesis, University of Cambridge. Abstract available online at https://www.repository.cam.ac.uk/handle/1810/241033.

87 Barbara Fruth and Gottfried Hohmann, 1996, 'Nest building behaviour in the great apes: The great leap forward?', in William C. McGrew, Linda F. Marchant, and Toshisada Nishida (eds), *Great Ape Societies*, Cambridge: Cambridge University Press, pp. 225–240.

88 Fiona Stewart, 2011, 'Brief Communication: Why sleep in a nest? Empirical testing of the function of simple shelters made by wild chimpanzees', *American Journal of Physical Anthropology* 146: 313–318.

89 For a discussion of Man the Hunted, see: Robert W. Sussman and Donna Hart, 2015, 'Modeling the past: The primatological approach', in W. Henke and Ian Tattersall (eds), *Handbook of Paleoanthropology*, Berlin: Springer-Verlag, pp. 791–815.

90 John A.J. Gowlett and Richard W. Wrangham, 2013, 'Earliest fire in Africa: Towards the convergence of archaeological evidence and the cooking hypothesis', *Azania: Archaeological Research in Africa* 48(1): 5–30.

91 J. Desmond Clark, 1993, 'African and Asian perspectives on the origins of modern humans', in Martin J. Aitken, Christopher B. Stringer, and Paul A. Mellars (eds), *The Origin of Modern Humans and the Impact of Chronometric Dating*, Princeton, NJ: Princeton University Press, pp. 148–178.

92 Anna K. Behrensmeyer and Kaye E. Reed, 2013, 'Reconstructing the habitats of Australopithecus: Paleoenvironments, site taphonomy, and faunas', in K.E. Reed, J.G. Fleagle, and R.E. Leakey (eds), *The Paleobiology of Australopithecus Vertebrate Paleobiology and Paleoanthropology*, Dordrecht: Springer, pp. 41–60.

93 John R. Gillis, 2012, *The Human Shore*, Chicago: University of Chicago Press, p. 29. People who occupy edges between ecosystems often show a greater resiliency. Gillis based his observation in relation to wind and water.

94 Richard Potts, 2013, 'Hominin evolution in settings of strong environmental variability', *Quaternary Science Reviews* 73: 1–13.

95 The need to take all types of environmental fluctuations is evident from the inconsistent trends found when deep-sea oxygen isotopes, terrestrial dust flux, paleosol carbon isotopes or lake sediments are considered together.

96 Richard Potts, 2013, 'Hominin evolution in settings of strong environmental variability', *Quaternary Science Reviews* 73: 1–13.

97 Mark A. Maslin and Beth Christensen, 2007, 'Tectonics, orbital forcing, global climate change, and human evolution in Africa: Introduction to the African paleoclimate special edition', *Journal of Human Evolution* 53: 443–464.
98 Richard Potts, 1988, 'Environmental hypotheses of hominin evolution', *Yearbook of Physical Anthropology* Suppl. 27: 93–136.
99 Ibid., Abstract.
100 Clayton R. Magill, Gail M. Ashley, and Katherine H. Freeman, 2013, 'Ecosystem variability and early human habitats in eastern Africa', *Proceedings of the National Academy of Sciences* 110(4): 1167–1174.
101 Richard Potts, 2013, 'Hominin evolution in settings of strong environmental variability', *Quaternary Science Reviews* 73: 1–13; Finlayson Clive, *et al.*, 2011, 'The Homo habitat niche: Using the avian fossil record to depict ecological characteristics of Palaeolithic Eurasian hominins', *Quaternary Science Reviews* 30: 1525–1532. Martin Ziegler, Margit H. Simon, Ian R. Hall, Stephen Barker, Chris Stringer, and Rainer Zahn, 2013, 'Development of Middle Stone Age innovation linked to rapid climate change', *Nature Communications* 4: 1905. Large variations in culture found in the African middle stone age, between 30,000 and 190,000 years ago, are attributed to behavioural flexibility; see Andrew W. Kandel, Michael Bolus, Knut Bretzke, Angela A. Bruch, Miriam Haidler, *et al.*, 2015, 'Increasing behavioral flexibility? An integrative macro-scale approach to understanding the Middle Stone Age of Southern Africa', *Journal of Archaeological Method and Theory* June: 1–46.
102 See Richard Potts, 2013, 'Hominin evolution in settings of strong environmental variability', *Quaternary Science Reviews* 73: 1–13. Note the abstract.
103 Mike Hulme, 2009, *Why We Disagree About Climate Change: Understanding Controversy, Inaction and Opportunity*, Cambridge: Cambridge University Press.
104 Ibid., p. 31.
105 Cited ibid., p. 31; and see A.J. Toynbee, 1934, *A Study of History, Vol. 2: The Genesis of Civilizations*, Oxford: Oxford University Press.
106 Discussed in H.J.B. Birks, 2005, 'Mind the gap: How open were European primeval forests?', *Trends in Ecology and Evolution* 20(4): 154–156.
107 G. Vines, 2002, 'Gladerunners', *New Scientist* 7 September: 35–37; K.J. Kirby, 2004, 'A model of a British forest-landscape driven by large herbivore activity', *Forestry* 77: 406–420; K.J. Kirby, *et al.*, 2004, 'Fresh woods and pastures new: From site-gardening to hands-off landscapes', *ECOS* 25: 26–33. John Birks, 2005, noted this point in the paper above.
108 Fraser J.G. Mitchell, 2005, 'How open were European primeval forests? Hypothesis testing using palaeoecological data', *Journal of Ecology* 93: 168–177.
109 H.J.B. Birks, 2005.
110 Lambros Malafouris, 2013, *How Things Shape the Mind: A Theory of Material Engagement*, Cambridge, MA: MIT Press.
111 Edwin Hutchins, 2008, 'The role of cultural practices in the emergence of modern human intelligence', *Philosophical Transactions of the Royal Society of London Series B* 363: 2011–2019. This delightful paper 'seeks the source of modern human cognition' and discusses the co-evolution of human culture and human brain, via way-finding in the Pacific Ocean to Greek orators. It concerns brain, body, and material and social worlds.
112 Epigenetics is the arena of research that deals with how genes are switched on and off, and what controls this, and when it occurs. This research is currently under great debate in genetics.
113 Environmental generational amnesia refers to the loss of knowledge with each generation if exposure is lost. Peter H. Kahn, 2002, 'Children's

affiliations with nature: Structure, development, and the problem of environmental generational amnesia', in P.H. Kahn and S.R. Kellert (eds), *Children and Nature: Psychological, Sociocultural, and Evolutionary Investigations*, Cambridge, MA: MIT Press, pp. 93–116.

114 The importance of the matriarch can also be seen in other mammals: elephants and more recently, orcas. Post-reproductive female orcas lead the groups in Canada during food-scarce years; both of these species are, like us, in the small global group of animals whose females live well beyond reproductive capacity. From Lauren J.N. Brent, Daniel W. Franks, Emma A. Foster, Kenneth C. Balcomb, Michael A. Cant, and Darren P. Croft, 2014, 'Ecological knowledge, leadership, and the evolution of menopause in killer whales', *Current Biology* 25(6): 746–750.

115 Bruce E. Wexler, 2010, 'Neuroplasticity, cultural evolution and cultural difference', *World Cultural Psychiatry Research Review* 2010 summer: 11–22.

116 Bruce E. Wexler, 2006, *Brain and Culture: Neurobiology, Ideology, and Social Change*, Cambridge, MA: MIT Press.

117 Wexler, 2010, p. 11.

118 Kevin N. Laland, John Odling-Smee, and Sean Myles, 2010, 'How culture shaped the human genome: Bringing genetics and the human sciences together', *Nature Reviews Genetics* 11: 137–148.

119 Ibid.

120 Of course, some of us cannot digest lactose, and this is geographically variable; Eugene E. Harris, 2015, *Ancestors in Our Genome: The New Science of Human Evolution*, New York: Oxford University Press, pp. 142–144. He also explains why some of us find Brussels sprouts bitter and some of us don't; it is in our genes, not due to someone's cooking.

121 The genome is made up of the genes and non-coding DNA and RNA.

122 Harris, 2015, p. 25.

123 We have more base-pairs than the house mouse, which has between 2.5 and 2.6 Gbase-pairs and just more than the rat's 2.75 Gbase-pairs. For a sobering discussion of the three mammals whose genomes have been decoded, see: Richard A. Gibbs, George M. Weinstock, Michael L. Metzker, Donna M. Muzny, Erica J. Sodergren, *et al.*, 2004, 'Genome sequence of the Brown Norway rat yields insights into mammalian evolution', *Nature* 428: 493–521.

124 To clarify the difference between genes and our genome: genes are the part of our genome that express characteristics, while the genome is the entire sequence of all base-pairs.

125 Malafouris, 2013, p. 241.

126 Bruce E. Wexler, 2006, *Brain and Culture: Neurobiology, Ideology, and Social Change*, Cambridge, MA: The MIT Press.

127 Ibid., 9.

128 Anna A. Adevi and Patrik Grahn, 2012, 'Preferences for landscapes: A matter of cultural determinants or innate reflexes that point to our evolutionary background?', *Landscape Research* 37(1): 27–49.

129 Differing generational experience relates very strongly with the ideas of environmental generational amnesia, and loss of connection with nature.

130 Giorgio A. Ascoli, 2015, *Trees of the Brain Roots of the Mind*, Cambridge, MA: MIT Press.

131 Renée Hetherington and Robert G.B. Reid, 2010, *The Climate Connection: Climate Change and Modern Human Evolution*, Cambridge: Cambridge University Press, p. 121.

132 Malafouris, 2013, p. 69.

133 Colin Renfrew made this comment in a discussion of cognitive archaeology: Colin Renfrew and Paul Bahn (eds), 2005, *Archaeology: The Key Concepts*, London: Routledge, p. 31. See also: C. Renfrew and P. Bahn, 2004, 'What did they think? Cognitive archaeology, art and religion', in C. Renfrew and P. Bahn (eds), *Archaeology: Theories, Methods and Practice*, 4th edn, London: Thames & Hudson. The great scholar André-Jean Festugière made a similar point (1954). He wrote that the historian has to rely on what he is told, but we can never penetrate the secret of the hearts of people who lived long ago: *La Révélation d'Hermès Trismégiste*, iv, Paris, p. 267.

134 Edward T. Jeremiah, 2012, *The Emergence of Reflexivity in Greek Language and Thought: From Homer to Plato and Beyond*. Philosophia antiqua, 129, Leiden: Brill, in Introduction, p. 1.

135 For a discussion of other Palaeolithic art see: Marcos García-Diez, Daniel Garrido, Dirk. L. Hoffmann, Paul B. Pettitt, Alistair W.G. Pike, and João Zilhão, 2015, 'The chronology of hand stencils in European Palaeolithic rock art: Implications of new U-series results from El Castillo Cave (Cantabria, Spain)', *Journal of Anthropological Sciences* 93: 1–18.

136 Melanie Pruvost, Rebecca Bellone, Norbert Benecke, Edson Sandoval-Castellanos, Michael Cieslak, *et al.*, 2011, 'Genotypes of predomestic horses match phenotypes painted in Paleolithic works of cave art', *Proceedings of the National Academy of Sciences* 108 (46): 18626–18630.

137 Jeremiah, 2012.

138 Ibid., p. 261.

139 Andrew W Kandel, Michael Bolus, Knut Bretzke, Angela A. Bruch, Miriam N. Haidle, Christine Hertler, and Michael Märker, 2016, 'Increasing behavioral flexibility? An integrative macro-scale approach to understanding the Middle Stone Age of southern Africa', *Journal of Archaeological Method Theory* 23(2): 623–668.

140 Deborah Bird Rose, 2004, *Reports from a Wild Country: Ethics for Decolonisation*, Sydney: University of New South Wales Press.

141 Malafouris, 2013, p. 5.

142 See entry on Process Philosophy in the *Stanford Encyclopedia*. 'Process philosophy' is essentially historically based and incorporates human experience as additive and changing. Process philosophy does not support a static position or a static psychology for our species.

143 William Davies, Dustin White, Mark Lewis and Chris Stringer, 2015, 'Evaluating the transitional mosaic: Frameworks of change from Neanderthals to Homo sapiens in eastern Europe', *Quaternary Science Reviews* 118: 211–242.

144 Ibid., p. 234.

145 Adherence to the theory as fact might well have restricted the questions to those that arise from the theory, a problem in science discussed by Don A. Driscoll and David B. Lindenmayer, 2012, 'Framework to improve the application of theory in ecology and conservation', *Ecological Monographs* 82(1), 129–147.

146 As discussed in John Hawks, 2013, 'Significance of Neandertal and Denisovan genomes in human evolution', *Annual Review of Anthropology* 42: 433–449.

Part II

THINKING ABOUT DESIGN

4

MULTIPLE, NOT SOLO VOICES

A father had a family of sons who were perpetually quarreling among themselves. When he failed to heal their disputes by his exhortations, he determined to give them a practical illustration of the evils of disunion; and for this purpose he one day told them to bring him a bundle of sticks. When they had done so, he placed the bundle of sticks into the hands of each of them in succession, and ordered them to break it in pieces. They tried with all their strength, and were not able to do it.

He next opened the bundle of sticks, took the sticks separately, one by one, and again put them into his sons' hands, upon which they broke them easily. He then addressed them in these words: 'My sons, if you are of one mind, and unite to assist each other, you will be as this bundle of sticks, uninjured by all the attempts of your enemies; but if you are divided among yourselves, you will be broken as easily as these sticks.'

The Fables of Aesop[1]

Who is this young woman in a canola field?[2] She is a landscape architect and is redesigning the family farm, something not thought of by designers twenty years ago. Landscape architecture final-year student Christie Stewart had the ambition to address large-scale environmental degradation in wheat-growing regions of the western half of Australia, a degradation characterised by rising and severe salinity, soil erosion, and loss of fertility. Christie's project was titled *Ameliorating Agriculture: Cultivating Biodiversity*. She used geographic information systems and an iterative design approach to test how new techniques of precision agriculture can be integrated with ecological systems design to enable a regenerating synergy between the agricultural and the non-agricultural. This synergy is key to the idea of working with multiple, not single voices, and heralds a profound change in how we treat the major ecologies of the farm, the bush, the veldt, the prairie, and the managed woodland or forest. This is about using scientific data to design for the best performance of interacting systems in landscape. In doing so in the realm of agriculture, landscape architects might truly be setting out on a new venture that could be styled *design georgics*, from the Greek γεωργικά – 'the things of the farmer'. *Design georgics* would lie on what Richard Weller has called the 'continuum of design intelligence'.[3]

Extraordinary changes are occurring in landscape architecture's relationships to agriculture. This chapter is about that change. Large-scale agriculture has been moving from an overriding premise of food production to considering biodiversity, management of water resources, the protection of soil structure and fertility for future productivity and carbon sequestration, greening corridors for connecting biodiversity and for ease of animal passage, and pockets of habitat for local endemic species. These imperatives need to be heard together, as one sound though many voices, with all voices sounding at any one time. I draw an analogy with counterpoint from music. Counterpoint is two or more melodies sounding together, where each voice maintains its distinct identity, but every voice has a harmonic relationship with all other voices.[4] Together, they sound in harmony. There is now a push for 'counterpoint' in many arenas in which landscape architecture deals.

The change to multiple voices on the farm has emerged through the twentieth century. The end-point ambition for the farm is what farmers desire above all – to pass on their land to future generations in the best shape possible for long-term futures. Old decisions we now see as erroneous were done in good faith based on the knowledge available at the time.

Here, I go beyond 'urban agriculture'[5] because vegetable gardens in urban areas will not solve the world's food crisis. To feed a projected

Who is this young woman in a canola field?[2] She is a landscape architect and is redesigning the family farm, something not thought of by designers twenty years ago. Landscape architecture final-year student Christie Stewart had the ambition to address large-scale environmental degradation in wheat-growing regions of the western half of Australia, a degradation characterised by rising and severe salinity, soil erosion, and loss of fertility. Christie's project was titled *Ameliorating Agriculture: Cultivating Biodiversity*. She used geographic information systems and an iterative design approach to test how new techniques of precision agriculture can be integrated with ecological systems design to enable a regenerating synergy between the agricultural and the non-agricultural. This synergy is key to the idea of working with multiple, not single voices, and heralds a profound change in how we treat the major ecologies of the farm, the bush, the veldt, the prairie, and the managed woodland or forest. This is about using scientific data to design for the best performance of interacting systems in landscape. In doing so in the realm of agriculture, landscape architects might truly be setting out on a new venture that could be styled *design georgics*, from the Greek *γεωργική* – 'the things of the farmer'. *Design georgics* would lie on what Richard Weller has called the 'continuum of design intelligence.'[3]

Extraordinary changes are occurring in landscape architecture's relationships to agriculture. This chapter is about that change. Large-scale agriculture has been moving from an overriding premise of food production to considering biodiversity, management of water resources, the protection of soil structure and fertility for future productivity and carbon sequestration, greening corridors for connecting biodiversity and for ease of animal passage, and pockets of habitat for local endemic species. These imperatives need to be heard together, as one sound though many voices, with all voices sounding at any one time. I draw an analogy with counterpoint from music. Counterpoint is two or more melodies sounding together, where each voice maintains its distinct identity, but every voice has a harmonic relationship with all other voices.[4] Together, they sound in harmony. There is now a push for 'counterpoint' in many areas in which landscape architecture deals.

The change to multiple voices on the farm has emerged through the twentieth century. The end-point ambition for the farm is what farmers desire above all – to pass on their land to future generations in the best shape possible for long-term futures. Old decisions we now see as erroneous were done in good faith based on the knowledge available at the time.

Here, I go beyond 'urban agriculture'[5] because vegetable gardens in urban areas will not solve the world's food crisis. To feed a projected

4

MULTIPLE, NOT SOLO VOICES

A father had a family of sons who were perpetually quarreling among themselves. When he failed to heal their disputes by his exhortations, he determined to give them a practical illustration of the evils of disunion; and for this purpose he one day told them to bring him a bundle of sticks. When they had done so, he placed the bundle of sticks into the hands of each of them in succession, and ordered them to break it in pieces. They tried with all their strength, and were not able to do it. He next opened the bundle of sticks, took the sticks separately, one by one, and again put them into his sons' hands, upon which they broke them easily. He then addressed them in these words: 'My sons, if you are of one mind, and unite to assist each other, you will be as this bundle of sticks, uninjured by all the attempts of your enemies; but if you are divided among yourselves, you will be broken as easily as these sticks.'

The Fables of Aesop[1]

population of 10 billion people, with increasing food protein aspirations and needs, we require 70 per cent more food by 2050. Urban agriculture at the small non-industrial scale in the city operates more in the realm of civic ecology. Exceptions do occur, such as the provision of up to 40 per cent of food production by the citizens of Havana following food embargoes through the 1960s until recently; this major and heroic undertaking has made Cuba a world leader in producing food in small urban gardens. Though there are lessons to learn from such food-growing systems, my ambition is to take landscape design away from small interventions for urban communities and into the bigger-picture performance of large-scale food production in rural communities. Larger-scale systems are a feature of the developed world and increasingly a feature of the developing world, though smaller landholdings currently make up a considerable percentage of world food production. Ironically, in many countries the poorest and hungriest people are those whose main occupation is farming.[6] In the developing world, food production might have to almost double.[7] The spotlight in this chapter is, therefore, agriculture, the main player in human food production.

A number of universities have 'rural studios' for undergraduates. Most of these are not dealing specifically with the production issues facing the agricultural sector but are focused upon social aspects of rural living in changing economic times. For example, in the School of Architecture at the University of Auburn's Rural Studio program, architecture students are designing homes for the low-income rural poor in west Alabama, one of the poorest regions of the USA. Similarly, the Rural Studio at the University of Western Australia focused on social-cultural exploration of life in a country town, including the cultural relationship with bushland, and the likely conditions coming with climate change.

In contrast to the social-rural studio, designers on the land and design students are working in a different manner. Rural Design, led by Dewey Thorbeck in Minnesota, is exploring what architecture can accomplish 'as a problem-solving process in shaping rural environments for the future', including some farming and environmental strategies to increase production.[8] All of these studios are valuable because, as Dewey Thorbeck points out, design schools and the design professions have long ignored rural regions.[9] Generally, rural studios do not address food or fibre production, the traditional realm of the agricultural scientist and farm adviser. However, changed imperatives within the farm now include issues outside the general remit of those disciplinary territories and suggest opportunities for designers. How might landscape architects contribute to the multiple voices of the farming landscape – food production, social engagement, connectivity of these constructed landscapes, biodiversity increases, water conservation or water use, and revitalisation of the landscape itself – whether that landscape is in agricultural production or a remnant of previous landscape?

What is agriculture?

Agriculture continues as the world's oldest and most significant constructed ecology and one of the oldest cultural activities in the world. This culture has irrevocably changed landscapes and fed the previous approximately 500–650 generations of humans. In a history of agriculture in the Cambridge World History Series, editors Graeme Barker and Candice Goucher note that approximately 10,000 years ago most of the world's population, of perhaps 6 million of us spread across the planet, were hunters, gatherers, and fisher-folk.[10]

As long as 21,000 years ago, wild barley was gathered and threshed on the shores of the Sea of Galilee;[11] this was not a planted crop, but was evidence of the massed gathering of food for group use. The human idea of collecting food survived the last Ice Age, but not until about 10,000 years ago did we begin to plant the food we needed. This was the 'agricultural revolution' of the Neolithic. For many years, and certainly when I was an agricultural science student some decades ago, agronomists presumed that the agricultural revolution arose due to better conditions for plants, herds, and humans when the cold conditions of the Ice Age ended about 11,700 years ago. This idea is a logical thought; after all, the winter temperature in southern France during the Ice Age went down to nearly minus 30°C;[12] temperatures in the Tigris and Euphrates valleys were likewise extremely cold with short growing seasons. We can easily imagine that the sun came out when the ice retreated, our forbearers all felt warmer and better, life got easier, more land was available, they discovered how to plant and thresh cereals, hunted more easily, and had more comfortable lives as a result of a newly expanded variety in our diet; that they had residual time and enough surplus food to build cities, trade, to invent pottery and its wheel, become political and philosophical, to invent writing, and have more children. This imagined scenario belongs to the idea that the elimination of stress causes progress. As I touched upon in Chapter 3, it is now considered that stress was the driver and stimulator of innovation in human history.[13] Indeed, the earliest examples of societies organised as states emerged at times of increasing aridity, such as occurred about 8,000 years ago.[14]

In keeping with new understanding of the connections between climate and agriculture,[15] many historians now consider that agriculture arose because, as the ice departed and the world warmed, there were changes in temperature and rainfall, with unpredictable weather; life became more and more difficult for most of our ancestors and they needed to change behaviours quickly.[16] Wild food plants shifted into novel areas, forests grew on pastures where they had easily hunted, and herds became widely dispersed, making it far harder for hunters to find meat. The end of the last Ice Age and subsequent climate shifts in aridity and rainfall made life difficult. These climate perturbations led to rethinking the art of human survival,[17] with experiments in the use of wild plants, herding, and sedentism, to increase the reliability of exploited resources. As a result, there are no generalisations about the rise of agriculture.[18]

Some regions, such as the 'Fertile Crescent' or the Indus Valley, are famed as centres for the origins of agriculture and the domestication of plants and animals. In the Old World the first clear indications of domesticated plants are found in south-western Asia by about 10,500–10,100 years ago, with evidence that cereals were domesticated by that time.[19] Main crops were emmer wheat (*Triticum turgidum* subsp. *dicoccum*), einkorn wheat (*T. monococcum* subsp. *monococcum*), barley (*Hordeum vulgare*); and the legumes lentil (*Lens culinaris*), pea (*Pisum sativum*), and chickpea (*Cicer arietinum*). Humans had brought sheep and goats under control by 10,000 years ago.[20] The food package of cereals, legumes, and meat spread across Asia, Europe, and most of the globe. With the need to record what was grown, and who sold what or was given to, came the invention of accounting and writing. It is a logical progression. While the connection is contested, that the Sumerian goddess Nidaba was the goddess of both grain and writing[21] suggests a tantalising link between these in the minds of those who lived at the time.

Tracking domestication shows how we moved from hunting and gathering to food production in village gardens, cultivated fields, orchards, and 'domesticated' forests. Manipulation of plants and animals is not just a key feature of the Holocene, but almost defines it. As an example, Figure 4.1[22] shows the domestication of crops in Amazonia, which was a world centre of plant domestication and included manioc, Brazil nut, sweet potato, cacao, pineapple, and tobacco. In Amazonia, a place still more usually associated with wilderness and hidden tribes, people began to domesticate plants and animals in the Early Holocene (by 8,000 years ago),[23] creating anthropogenic soils in more densely populated areas;[24] humans appear to have been key agents of disturbance during this long history and lived through changing climatic and forest conditions.[25] In 1492 when Columbus arrived in the Americas, few pristine landscapes existed in Amazonia; the population then was about 8 million; it is approximately 34 million now.

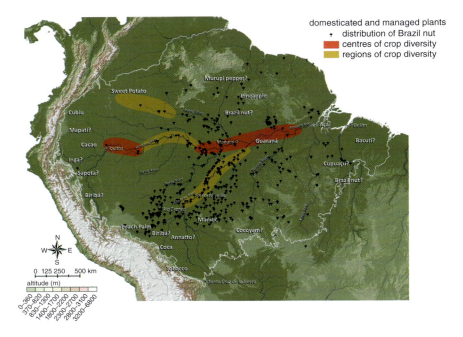

Figure 4.1 Plant management and domestication in Amazonia, showing the distribution of the Brazil nut. The names of species identify known or suspected (with '?') origins of domestication of twenty native Amazonian crop species. The focus is along rivers, with scattered settlements between.

Source: Image from Clement, *et al.*, 2015, 'The domestication of Amazonia before European conquest', *Proceedings of the Royal Society B* 282: doi: 10.1098/rspb.2015.0813, with permission from the Royal Society.

Today agriculture is again at a major crossroads and experiencing a crisis in how we farm. What is emerging is a reassessment and revitalisation of the agricultural sector, again a sign of *shifting adaptabilities* of the human mind and spirit. This situation is not new for food production, even for comparatively recent times. For example, in medieval western Europe there was a similar shift in approaches to farming that led to an important but not widely known agricultural revolution. That medieval revolution focused on the introduction of fodder crops and crop rotations, convertible husbandry, and the introduction of clovers, turnips, and other root crops.[26] This revolution really ended its period of intense innovation only 150 years ago, in the middle of the nineteenth century, and set the stage for the modern world's highly programmed agricultural production.

More widely known is the 'green revolution' that began in the 1950s.[27] Its focus was on replacing old varieties with new ones, and the revolution was applauded due to the increased yields that were achieved. But yields have not been sustained, and many small-scale farmers in the developing world did not benefit much. In addition, that revolution of short-lived prosperity has left us with the problem today of the loss of crop genetic diversity, an unanticipated implication of the extensive global push towards cropping monocultures of new varieties.[28] It is only in remote regions, where the 'modern' world (including aid agencies) failed to reach, that individual fields can still be found packed with enormous genetic diversity.[29] This genetic material now provides us with molecular means to improve food production far more quickly than with traditional crop-breeding techniques.

Today there are three essential and interconnected issues of food security – loss of biological diversity of crop species, responses to emerging climatic changes, and pressures on our continued ability to ensure food supplies. Because the farm is no longer seen as an isolated island of cultivation, but part of the wider ecosystem, farmers are now, in a new revolution, addressing many other issues to preserve soils, water, and natural landscapes. Many farmers, farming with caution, have carried out practices along these lines for centuries.

Climate change will exacerbate the challenges of food production. While designers are more often concerned with the impacts of climate change on human living spaces, notably cities, some of the most profound impacts of climate change will be on our food systems. Climate change will test our abilities to change and modify current ideas.[30] It is important that in the rush for designers and design schools to engage with the urban, and hearing the numbingly repeated information that half of the world's population now live in cities – and that designers should therefore focus on cities, we must not lose sight of the fact that all of those city-dwellers need to eat and that the other half of humanity do not live in cities. For example, approximately 1.4 billion people live in subsistence villages of substantial poverty. Bearing in mind the problems of urbanisation, it might be good to assist villagers to not move to cities, to create Smart Villages to match that of Smart Cities,[31] and to improve the life and economic potentials in rural farming communities. The organisation Smart Villages focuses on remote off-grid villages; the first aim is for access to sustainable energy, as a catalyst to food security and clean water, healthcare and better nutrition, improved sanitation, education opportunities, business opportunities, and productive enterprise to increase village incomes, gender equality, and democratic engagement.[32] The essential quest is to keep farmers on farms.

Keeping farmers on the land is a prime concern today.[33] One of the great urgencies is the worrying move away from farming by youth in many countries, whether developed or undeveloped. For example, in the United States the average age of a farmer is nearly 60, while in Africa younger people show reduced interest in farming. Wherever in the world we look, agriculture is seen as hard work, continually hit with economic and environmental vagaries, and many consider it 'old' – old-fashioned, not vibrant, and not modern. Indian cities attract villagers due to their technology and the allure of the 'modern' and the promise of lifestyle variability. However, new initiatives are working to make a 'frugal innovation revolution'[34] by which it is hoped to enable sustainable energy systems that will, in the first instance, work towards improving lives in rural areas. The very 'idea' of agriculture needs to change in order to address the loss of farmers, the mental health needs of farmers, loss of economic opportunities in rural areas, rural-urban tensions, changes in rainfall and temperature with climate change, and changes in natural resources on the land.[35] While the social-cultural is the arena where landscape architecture's current Rural Studios concentrate, we can do more in the agricultural arena. It is in this realm of shifting culture and changing environments that landscape architects might find a strong role by providing new design ideas in order to help transform the perception of agriculture to one that has a strong and exciting future of many opportunities.

Multiple and explicit voices in agriculture

Many new voices now need to be heard in agriculture in addition to food and fibre production. These additions are emerging from better knowledge and understanding of biological, chemical, and physical processes. In Table 4.1, I have set out an estimation of some of the voices that have emerged in large-scale farms of the developed world. A table like this will vary with region, products grown – whether animal or plant or both – cultural history of farming practices, cultural history of communities, politics, and finance. Emerging changes in agriculture are enormously diverse around the globe, with only a few universals (such as soil preservation). For example, in some parts of the world, 'living with predators' will be included – the tiger in India and the wolf in the USA. However, from revisions of hill farming in Scotland, regeneration of poor village lives in southern Asia, or large-scale ranching in the United States, times are changing on the farm. Working with multiple voices is not confined to the developed world or to large-scale agriculture. In the developing world, working with indigenous knowledge of the local environment, using modern agricultural science, and exploring diverse crops to combat uncertainty suggest huge opportunities to improve farming practices.

Table 4.1 Changing imperatives, or voices, on the farm: a generalised view of best practices and their evolution. This list will change greatly for every region and local area (this one is geared to Australia), but is an indication only of the mix of growing imperatives, and how these are widely distributed across various knowledge landscapes. It also suggests where landscape architects who are well versed in ecological complexities could make a contribution alongside colleagues in the sciences.

Imperatives on the farm	1900–1910	1920–1930	1940–1950	1960–1970	1980–1990	2000–2010
Food and fibre 1900	→	→	→	→	→	→
Soil science and plant nutrition 1900	→	→	→	→	→	→
Removal of 'vermin' animals 1900	→	→	→	→	→	→
Water management 1940s			→	→	→	→
Increased awareness of overstocking impacts 1970s				→	→	→
New cropping ideas to conserve soils 1970s				→	→	→
Indigenous values (First peoples) 1970s				→	→	→
Biological conservation 1990s					→	→
Awareness of farm connectivity to natural areas 1990s					→	→
Soils restoration 1990s					→	→
Preserving wildlife through feral animal control 2000						→
Ecological bushland corridors: active measures being taken 2000						→
Farm as ecotourism 2000						→
Maintaining farmer mental health 2000						→
Soils emergency 2010s						→
Carbon farming 2010s						→
Ecosystems for climate buffering (carbon sequestration) 2010s						→
Genetic diversity of crops 2010s						→

Soils, a major voice in all agricultural systems

Soils are basis of all health and life on terrestrial Earth. It is a shocking thought that we are 'running out of soil' all over the world. This means that soils are being depleted not just of their nutrients but of their microorganisms, because soils need to be microbiologically alive to give us plant production. The importance of soil microbiota (bacteria, actinomycetes, and fungi) and macro-biota (such as earthworms) is often not understood. In the 2015 movie *The Martian*, a stranded astronaut grows potatoes in lifeless Martian soil. In reality, he would have little chance of much of a crop, as microbes are needed to convert organic material such as manure into inorganic molecules that a plant can absorb.[36] The soil community affects the productivity of the plant community above it.[37] This all highlights the simple point that the ecosystems of the one planet we have are intricately bound by the evolved histories of all its biological, chemical, and physical components. The movie also served to show that we still know less about Earth's soil than an alien planet. That this point was noted by Leonardo da Vinci[38] shows a remarkable stasis in knowledge about the foundation of all our food.

The preservation and restoration of the largely hidden world of soils is a major voice on the farm and has been growing louder. Soils are core to new ideas of regenerative or transformative agriculture. Transformative agriculture for soils has been focusing on a number of practices, such as reduced tillage, biochar (charcoal produced from plant matter and stored in the soil as a means of removing carbon dioxide from the atmosphere), improved crop rotations (continuing the development begun in the Middle Ages), better nutrient management, cover crops (that protect and enrich soils), land restoration, and agroforestry. Increasingly, these require data management, research-measurement networks, soil-monitoring networks, greenhouse-gas data management and networks, remote sensing, and model testing.[39] All of these point to a major shift in the way the farm will be operating in future, with opportunities for decision-support from designers, should we take up this challenge.

A voice: Biological conservation on the farm

The distribution of more than half of the world's plant and animal species lies outside protected areas such as reserves and national parks, and lies within primarily agricultural landscapes.[40] Yet, as noted by David Norton and Nick Reid, authors of *Nature and Farming: Sustaining Native Biodiversity in Agricultural Lands*, only a fraction of original biota is able to survive in intensively developed agricultural landscapes, such as cropping areas, and even less in irrigated agriculture. Farms offer a spectrum of agricultural landscapes to provide opportunities for biological conservation. Where a farm lies on the spectrum determines the degree or type of intervention by which designers can improve biodiversity. The spectrum of opportunity moves from rangeland grazing, or pastoralism, which is extensive agriculture, to intensive forms – such as high-intensity grazing, and intensive cropping systems – and to very intense irrigated cropping and pasture and horticulture. Low-input extensive agricultural systems are far more compatible with the maintenance of native biodiversity than are intensive systems because intensive systems have usually removed the majority of pre-existing vegetation. These situations present particular challenges, but within the spectrum are new horizons for designers to assist biological conservation.

To compound the spectrum differences between extensive and intensive agriculture has been the rise of techniques and technologies (such as water extraction) to farm more and more land that had once been considered too marginal to farm. In the developed world, this move was partly a response to the cost-price squeeze faced by farmers since the 1950s.[41] At the same time, some faced crippling economic conditions that led them to abandon pastoral land. Abandonment has led to opportunities for land to recover from being stocked (as in Figure 4.2) and has led to opportunities for wildlife preservation and wildlife-friendly farming.

Figure 4.2 Wildflowers in the rangelands of the Murchison region, Western Australia, late winter 2015, after good rains. This pastoral area carried sheep for wool until recent large-scale de-stocking in response to changing local conditions, and is now primarily grazed by a few meat sheep, cattle, rangeland (i.e. managed) goats, kangaroos, emus, and feral donkeys and rabbits. Major changes are occurring in soils and biodiversity. All over the world, managers reimagine ecologically sustainable management of rangelands.

Image: With permission, Karen Morrissey.

A voice: Crop biodiversity and climate spaces for different climate years

While designers are concerned with maintaining biodiversity and ecosystem function in revegetation and rehabilitation, one of the most serious components of biodiversity loss has been in food crops.[42] Agro-biodiversity takes us once more into the realm of discussing 'genetics for designers', as in Chapter 2. Most of human life today is supported by only twelve plant species – rice, wheat (which takes up 17 per cent of the world's cultivated land), maize, millet, sorghum, potato, sweet potato, soybean, sugar, cassava, plantains, and yams; these provide 75 per cent of the world's food. The three mega-crops – rice, wheat, and maize – provide 50 per cent of the world's plant-derived energy. Fifteen crop species provide 90 per cent of human energy.

Without the assistance or benefits of modern agricultural science, poor farmers on small landholdings produce one-fifth of the world's food.[43] The majority of resource-poor farmers are women, and their produce includes cassava, groundnuts, pigeon peas, lentils, cowpeas, yams, bananas, plantain, and wild species of vegetables.[44]

There are approximately 23,000 edible plants on the planet; we eat about 300 of them.[45] World focus on only a handful of species is a remarkable position, and over the last 100 years research has worked hard to improve these select few to the near exclusion of all others. 'Orphan crop species', often local foods or old versions of a more widely grown crop, have been forgotten about or neglected. Yet in response to climate change, many of these orphans have tremendous potential to be improved with a modicum of research.[46,47] An example of this unexploited 'treasure trove' of crop diversity that needs research is the Bambara groundnut of West Africa, an indigenous legume that is drought tolerant and a good performer on poor soils.[48] New crops such as this will assist us to face drying conditions and provide options for farmers that will help keep them on the land, and keep us all fed.

The performance capacity of crops in different types of climate years is critically important. We need to maintain both crop genetic diversity and species diversity in pastures and connected, natural landscapes because some species specialise or flourish in particular climate years, while others decline or fail. In old agricultural fields, farmers traditionally kept a mixed sward of varieties, knowing from experience that an unexpected weather event, such as a longer-term drought, will put performance pressure on the crop. By keeping a mix of cultivars or varieties in any single field, they traditionally kept open the possibility that at least part of the crop would cope with whatever conditions occurred in the growing season. While many cultivars can co-exist through most seasons, the performance of a field in the high-pressure season of difficulty is crucial for maintaining some crop.

Landraces – 'nature's pedigree seeds'[49] – are the varieties that generations of farmers have selected and sown year after year because they consistently perform well in their local environment. Performance drives the search for landraces, and the need for preservation of these is key to gaining genetic material for resilience to climate change. Landrace preservation is important because it gives us genetic diversity that translates into a more complex suite of opportunities for future food.

Landscape architects who are excited by the ideas of performance, GIS, and the spatial articulation that those techniques give, and can translate those into cropping systems, would be well placed to assist spatial mapping and testing of new crops, old crops, and old cultivars. Within this testing we might well be playing with roughness, topography,[50] and the idea of long-tailing discussed in Chapter 2's climate space, and the articulation and design of best fit. In this disciplinary scenario, the designer would need to work closely with soils and meteorological data, and a host of unknown parameters, alongside agriculturalists. Agricultural landscapes occupy about 40 per cent of the terrestrial Earth,[51] suggesting that a more serious involvement from landscape architecture is compelling.

A voice: Non-crop species diversity within agricultural and connected landscapes

While agricultural ecosystems are habitat for many species, there has overall been a negative impact, especially with the intensification of agriculture and the move to large-scale farm holdings and landscape simplification into larger and fewer fields. The increased use of pesticides and fertilisers have also reduced biodiversity in agricultural landscapes. However, rather than focus on this oft-told negative story of the ills of agriculture, it is well to note that there is a wide range of initiatives all over the world that are working out how agriculture might support species diversity. With a big-picture view, landscape architects can contribute to this ambition by developing methods to test how non-crop local plants might respond spatially to climate years, working with plant physiologists and climate scientists at the landscape scale. Are there ways in which we can assist the persistence of less frequent species beyond the provision of simply heterogeneous environments?[52] To do this we might well take some of the ideas of second nature of the French landscape architect Giles Clement out of the garden and into the *design georgics*. Added to ideas of second nature will be the long intelligence of the science of agronomy and knowledge of how species mixes change over time and space, and how seeds disperse.

Design voices within agriculture

It is my belief that landscape architects can have a deeper role in agriculture, and here I set out how, and in what capacities. I discuss the idea of *design georgics*, and the more precise role which landscape architects might play in agriculture.

Design voices: Climate space in agriculture for landscape architects

A challenge will be that individual species, plant assemblages, 'hotspots', and national parks will all be migrating with climate change in the coming centuries and will be re-sorted and reshuffled, as outlined in Chapter 2. These changes will need to be mapped, if mapping is possible, or speculations made for maximum likelihoods of where plants and animals need to go. In many regions, farm landscapes will be essential sites in maintaining biodiversity and connectivity between landscapes at both the species and population levels, and in creating climate spaces for regional species. An important issue, as noted decades ago by the distinguished environmental historian Donald Worster,[53] is that environmental boundaries are rarely the same as political boundaries.[54] How might we assist the challenge of biodiversity preservation across national, state, regional, and local governance boundaries? Connectivity and passages for movement need to be designed at a large scale. While great difficulties remain for determining where we might create climate spaces for future centuries to deal with reshuffled climate conditions and species, design always works towards a defined end in the decision process. Because of this, landscape architects can add to the profoundly important ecological need and theoretical discussion of where species might be going.[55]

As noted in Chapter 2, landscape architect Richard Weller and associates at the University of Pennsylvania are mapping to reveal the differences between targets for protected habitat given by the United Nations and the reality on the ground. Their purpose is to give some direction towards the extension of biodiversity regions, with a view to maintain ecological networks for biodiversity. Their *Atlas for the End of the World* project will enable community groups and farmers to pinpoint routes of best outcomes to focus their money and effort. Working with the woods, forests, wetlands, seagrass beds, mangroves, scrub, and heathlands that we have, and extending or developing them with precision, is important because intact ecosystems are the best buffer against climate change. We do not need technological fixes or complicated design.[56]

Important in this discussion are the ideas of land-sharing and land-sparing. Land-sharing is an idea from conservation and is concerned with producing both food and wildlife in the same landscape, with the ambition to aid wildlife by adding natural elements in the farm. However, land-sharing has led to reduced yields, and it seems that trying to do both in one space – all mixed up – is not the best answer. Land-sparing, an idea from agricultural science, focuses on increasing yields on farms while sparing remnant habitat from farming and restoring land that is freed by more productive agricultural land. Evidence largely supports land-sparing as being more successful for both food production and wildlife habitat.[57]

Land-sparing, where design might assist precise land use from information from both farming and conservation, is a type of design work that will require large amounts of data.[58] Increased data is of growing concern in farm management, and increased data should be of interest to landscape architects pushing the boundaries of the profession. How can designers use their data-capacities to assist farmers?

Climate space interventions occur at other scales. A small example is *grass-bank farming*, an initiative in the USA to assist when feed has been failing due to poor seasons. A grass-bank is exactly that – a bank for grass, for a farmer to draw upon in a difficult time; it is an easement through which grasslands are conserved and restored while enhancing ranching or pastoralist opportunities.[59] Such organisation structure is perfect for small landholdings and is, at the same time, a small adaptation towards creating a type of climate space for pastures. Crops will require similar initiatives.

Regenerative agriculture

Designers in the firm Nelson Byrd Woltz, who run a Conservation Agriculture Studio, believe that landscape architects are capable of engaging in regenerative agriculture, which is a form of restoration ecology and proactive conservation in an agricultural setting. Orongo Station, a 3,000-acre sheep farm on the east coast of the North Island in New Zealand, is a site designed by Nelson Byrd Woltz with the ambitions of regeneration and stewardship. They carried out this project at the time scale of agriculture's long views – a decade – and involved multiple voices – sheep farming, pasture rehabilitation, expansion of a Maori burial site, building restoration, reforestation, reconfigured wetlands, and orchards (Figure 4.3).

Nelson Byrd Woltz worked on their rural projects with conservation biologists, soil scientists, farmers, farm managers and advisers, and landscape ecologists. While agrarian landscapes are an opportunity for landscape architects, Thomas Woltz considers that this arena is often seen as outside our remit.[60] Yet, as conservation biologist James Gibbs noted, scientists 'can define the biological limits' and landscape

Figure 4.3 Orongo Station, Poverty Bay, New Zealand; designed by Nelson Byrd Woltz 2001–2012. The images show wetland construction (top), wetland and orchard (middle) and the alignment of allées for wind protection, with a 300-year-old *Ngai Tamanuhiri*, a Maori people burial ground, seen in the distance (bottom).

Images: With permission, Nelson Byrd Woltz Landscape Architects.

architects 'can define the possibilities of design'.[61] Here again are Lewis Mumford's *restrictive conditions* to assist us in the act of design and extend the profession in critical and specific ways.

Regenerative farming is also occurring with the recultivation of previously abandoned land, notably in temperate areas where land abandonment occurred due to agricultural intensification. An example is the black earths of south European Russia, where young emerging forests on abandoned land work as carbon sinks. We still need to study how these regions might be redeveloped because they are sites of a great number of competing issues;[62] yet many of these voices need to meet to enable both survival of communities and preservation of ecosystems. Many regenerative or transformative ideas, whether large or small, work with accepting change and seeking innovation, and sometimes work with a return to past knowledge, or draw upon local and traditional techniques based on precaution.

Carbon farming is also part of the regenerative and transformative approach to agriculture today. The idea that carbon farming is a way of assisting natural revegetation and the restoration of biodiversity in agricultural landscapes is a comparatively new one;[63] however, it is set to grow depending on the region.

The use of data and roles for designers beyond restoration agriculture

Farmers in the developed world, now being confronted by Big Data, need assistance on how to manage data, and how to manage the gap between data and decision-making on the farm. Yet farmers have little time to deal with these particular complexities and are already overrun with economic and meteorological data. The busy-ness of farmers suggests a new opportunity for what have been termed third-part intermediaries,[64] who can provide decision-support to farmers using Big Data. This opportunity suggests a changing world of the farm. Whoever might be working with the farmer needs to engage very specifically with a farmer's concerns, local nuances of climate, soil, water, biodiversity, carbon, and local and national economies. The gap between data and decision-making points very strongly to a new role for a new generation of graduates who can work between landscape in its widest sense, and who can work with data from agronomy, soils, and business, using data from various systems in the landscape to redesign the farm. Here we have the potential for landscape architects to be working in the arena of decision-support in the agricultural sector. Such interpretive practitioners will be essential; they are not available yet, but will come because farmers must be able to compete with markets to get the most out of their land without jeopardising its future productivity or environmental resilience and sustainability.[65]

The future scenario outlined here of a more explicit involvement in farming is going beyond 'restoration agriculture'. I am suggesting that landscape architects can be far more actively engaged with data use, scenario-building, and the generation of iterations of possibilities that many farmers are now struggling to generate without assistance. A more active engagement leads landscape architecture to the realm of precision agriculture that could be developed in many agricultural systems.

While traditional agriculture is practised according to generally predetermined schedules, precision agriculture employs a suite of real-time data on weather and air quality (humidity), soils, crop maturity, and issues of labour and machinery hire and use. Precision agriculture uses predictive analytics to make best utilisation of scarce and precious resources to maximise production with least damage to the environment. This is a very smart way of working and uses sensors in crops. Precision agriculture is a revolution in agriculture. It joins strongly with the abilities of new generation design students to work with data[66] because there is little difference in this 'farm work' and testing for temperature, pollution, wetland performance, or tree canopy impacts in a city. Designers who engage or are interested in the latter projects would be able to assist on the farm. Currently, due to the requirements of information technology infrastructure, monitoring, and the specialist nature of this work, precision agriculture is primarily undertaken by large companies. However, this monopoly is predicted to change rapidly[67] because precision agriculture can be a game changer for farm efficiencies, as farmers face tougher conditions and falling prices for their commodities. Precise new techniques and fine-tunings will lower energy use, place fertilisers more precisely, plant just the right amount of seed, prevent overlapping of spray, and increase overall yields in a more sustainable way. These technologies will require a new type of landscape practitioner to assist the global ambition of increased food production.

An important aspect of Big Data is that, as a non-rival good, it does not depreciate if more people 'consume' it.[68] Many can partake and jobs will open; here perhaps lies that mythical beast – the 'as yet unimagined job'. Great potential exists for landscape architects to work with agronomy, Big Data, ecological knowledge, and biodiversity across farm landscapes. Here, we would be working at perhaps the largest scale that designers are likely to work, particularly if we chose to work in large-scale agriculture, pastoral country, or rangelands. The landscape architect Jane Amidon recently called for landscape architects to be 'data-savvy' and to work at large scales commonly associated with planners.[69] In taking on the data challenges presented by the core demands of agriculture, designers would be stepping out, tackling head-on a major world issue, and leading an experiment in what it means to be a designer. While architects all over the world are engaging more and more with landscapes and not just the buildings, *design georgics* is one arena where the landscape architect could be ahead of the game.

Small holdings make up a third of the world's agriculture. They also provide the same opportunities for assistance, though many will occur in regions of low data support and of absent data and information. However, working with the challenge of low data support would be particularly attractive to many young designers, and I address some of the issues of missing data in Chapter 6.

The use of sensors and systems on the farm is at a distance from many of the high-profile, high-technology solutions to the problem of food production that are currently flourishing in design schools. Many of these urban systems are hydroponic – with high requirements for water, energy, light, and plastic parts – and produce very little food. Many appear to be rediscovering the long-held findings of experimental agricultural science. Ultimately, these high-profile systems often appear as intellectual and physical toys for urban designers. While these systems might be economically, ecologically, and socially viable in some settings, such as the mega-cities and certain affluence levels, they will not provide the volume required to solve the real problem of world food. The harder task is before us, and technology needs to be directed to work within existing farming systems and farming societies to make them workable, successful, and climate-resilient, and to give the next generation of rural communities a socially and ecologically sound future. New high-tech suggestions for city-based agriculture take us away from ecological processes, chemical and physical realities, and presume that energy for robotic systems, water, and lighting are always available, when such technology is neither reliable nor existent for much of the developed or developing world. High-tech urban food-growing also suggests an isolation of the city from the farm, a distancing of the urban from the rural, and an indifference to the realities of farming that arise when food comes in a package under a fluorescent light. If a vegetable patch fails in these high-tech systems, there is always the supermarket nearby. Such unreality is not what agriculture has ever been about. The very word comes from the Latin *agri cultura*, the cultivation of the land, and the Latin also refers to the country, as opposed to the town.[70]

Some constructions in the city claim that they are producing cleaner food, a surprising claim given the air pollution within many large cities. The greater challenge to us is to clean contaminated agricultural land, to listen to the voices of farmers and give them support, heed the wisdom and knowledge of farmers and agricultural scientists, and not produce glossy magazine images of urban greening under the guise of food solutions as seen in the minds of the city-based design professions. Landscape architects who work with farmers will also be rejecting the urban myth that in the city we can address everything over a café latte – 'urban agriculture' will not solve current food crises. The city is not the solution to everything.

Concluding comments on designs for the farm

Farmers and rural communities are tackling the massive challenge of juggling ambitions upon which all humanity depends – looking after the food, the land, the water, the atmosphere and the wildlife, and much of the economy.[71] As part of thinking of these landscapes in terms of multiple voices in counterpoint, I note that the various voices present, coming and needed, will change, fluxing with varying conditions, dropping out and coming back if needed. Foci will change over time and in response to environmental, economic, and climatic conditions from year to year and decade to decade. There are no solo voices, because in order for farms to have long-term futures they need maximum flexibility and resilience. The challenge for designers to assist persists and increases. Assistance will not come easily to us but will extend our thinking and expand our design capacities.

Other issues for design and food

In a chapter centred on agriculture it is fitting to address the voice of 'urban agriculture' in public open space in the city. This topic appears to be a popular form of easy greening by design students and others, with little thought or understanding that the agricultural and horticultural sciences are major arenas of intense inquiry and experimentation, with long histories that can be drawn upon.

Designers need to be aware of exactly what their ambition and role is in 'urban agriculture'. Too often urban agriculture is given as a greenwash or as a walk-through green experience, particularly within architecture, and is a facile gesture, not substance. But what is 'urban agriculture', and what is it really achieving? Place large vegetable planter boxes in a design, and you have agriculture. This is not the case. Let us not be naïve as design disciplines and let us know a scheme's limits, and the limited capacity of the urban environment to provide food. Many of the designs that I have seen are really token interventions of urban greening and community involvement in greening, or growing plants in bespoke scenarios, not about the reality of producing food at economies of scale. Most, if not all, food grown under the auspices of 'urban agriculture' does not consist of the major food crops of the world that support millions through export and large-scale production. A modern city could not produce enough wheat to make bread or enough rice to feed many. Designers need more acuity in articulating, with some modesty, their or the community ambitions for urban agriculture. Urban agriculture is less to do with food production and more to do with social amenity and civic greening in most developed cities, including the desires for 'cleaner' food with less food miles for those who can afford it. This will appear a contentious or even inflammatory statement to some people, but my training as an agricultural scientist gives me concern that 'urban agriculture' is actually something other than what its name states, and the name is deceiving.

Far closer and better descriptions can be encapsulated in the ideas of civic ecology,[72] and civic agriculture.[73] Human contact with soil and plants is good and healthy for the human spirit, but it will not feed the world, whether that world is the 'first', 'second', 'third world', or the so-called 'global north' or 'global south'. If we go back to the lists of the world's great crop species, we can see that few if any can be grown within urban landscapes at the economies of scale needed to feed the world; nor are we likely to succeed in doing so because such a quest would be to defy the entire human history of agriculture. Civic agriculture is really about urban horticulture. It is good for children to appreciate that it is fun to grow corn, that quite often fruit and vegetables are attacked by insects and large caterpillars that munch them, that plants suffer diseases, that urban rats also enjoy collecting your tomatoes, that birds often eat your ripe fruit the morning you were going to pick them because the birds got up earlier, and that growing enough food to live on would not be and is not an easy exercise. But can 'civic agriculture' be more than that?

Some 'urban agriculture' initiatives go beyond greening and vegetables into environmental and nutritional education in community gardens; this is primarily a social initiative. These schemes are being done with agri-food advisers and have the specific aim to improve food infrastructure and local food availability at some sort of economy of scale, and with a view to increase the availability of nutritious food to underprivileged people in low-income areas where nutrition is poor. Such schemes are more and more prevalent. In Ontario, for example, Karen Landman of the University of Guelph[74] has a long-standing interest in food deserts – where people have little access to fresh food – and in these cases urban horticulture gives benefits by providing fresh vegetables to people who often have little access to any but canned food, with serious health impacts. The landscape design firm Stoss, working with a non-profit called Seed Capital Kentucky, has designed a 24-acre site for food production, processing, food education, food training, and social activity; it includes job training and entrepreneurial programmes. As Chris Reed, the principal of

Stoss notes, they are working 'far beyond the scope of the designer'.[75] These are really urban fresh food centres in public spaces, and a great deal of research remains to be done as to how such projects are best delivered, and what they deliver. Many new initiatives have come from depressed social conditions.[76]

In many ways, the popular move to urban horticulture might be considered to be part of the long continuum of how we have developed food production and of our relationship with food crops. With urban horticulture, urban workers are perhaps attempting to increase the reliability of food resources, and to decrease the stress of urban living and the extinction of experience of the natural world within the modern city. Growing one's own, or providing local green supplies, fulfils a need to ensure quality, nearness, and freshness. However, designers need to be aware of the reality of what is actually being achieved, and to be explicit about their objectives.

There are three central issues that 'urban agriculture' does not address. First, every piece of land used for 'urban agriculture' denies land that might have provided for local regional biodiversity, habitat provision for local species, or improved connectivity of green spaces in urban and suburban areas. 'Urban agriculture' is in conflict with these voices within the city. A second problem of urban agriculture is that it is not drawing attention to the inexorable loss of farming land near cities due to urban sprawl. We have been building on our best soils, and they are then covered under impervious surfaces and lost, with devastating impacts on the life within the soil. Third, it could be claimed that urban agriculture is drawing attention away from the major problem of agricultural soils loss by suggesting subtly that it is in the city that we can solve agricultural production, or somehow make amends for lost agricultural land, or counteract loss by clever technology. Such thinking is taking us further away from addressing the key issues of lost agricultural land and food productivity, and further divorcing the city from the country – that old problematic divide of humanity. Food production cannot be the role of the city; even the recent concept of land-sparing teaches us that. This division and blindness are major follies. Current thinking in many design schools needs a complete refocus to address the more difficult – rural agricultural production.

Loss of good agricultural land is of far greater crisis and importance than 'urban agriculture' in the developed world and the rapidly developing world, which is, with sprawling cities, evicting working farms on good soils and replacing them with high-rise flats. In Chinese cities the destruction of near-urban soils and the displacement of the farming communities that have worked that land for thousands of years raise enormous and vocal discussion from my Chinese national students, who see in the destruction of those farms the destruction of their deepest cultural roots. Such emotion and intellectual concern are reminders to us that the science and art of agriculture is one of our very oldest, most important, and long-cherished cultures.

Conclusions

Many arenas of design and ecology today have been steadily moving from mainly single-voice considerations to the need for multiple voices. I have focused on agriculture. National Parks are another case. One hundred years ago, in 1916, the United States established the first national parks, and this has been claimed, perhaps cheekily, as 'America's best idea'.[77] At the time, managers saw national parks as formal places with the defined purpose of providing for recreational tourism.[78] Wildfires were 'routinely suppressed … to protect the scenery, wolves were eradicated', and bears were fed to create a spectacle for tourists.[79] Over the course of the twentieth century, biologists and forest managers added to and altered the voices needing to be heard in national parks[80] – fire prevention to preserve the spectacle of the national park changed to fire accommodation as they understood the importance of fire, particularly in some global environments such as the eucalyptus forests; and in the United States managers ceased pesticide use to control insects,[81] and reintroduced wolves. All new strategies suggest a fine-tuning, with a multiple-voiced approach to working with landscapes. Farms and national parks are now no longer seen by their managers as merely blocks of land or islands of decision-making with no recourse to the wider ecological world. We have moved from a world of limited or single voices to the consideration of multiple voices.

And yet researchers do far too much research with a focus on only one voice, one line of inquiry, or one disciplinary lens. Single issues that fail to address reality, complexity, and uncertainty have often led to skewed views because research rarely gets to examine the part in terms of the whole.[82] Piecemeal thinking can mislead us both in understanding and in design outcomes, and push us astray for years if not generations; piecemeal thinking is too easy, with an expedient result, but not a true or complete answer to the questions challenging us. Single issue analysis does not give us the multiple-part strength seen in Aesop's fable of the bundle of sticks.[83]

Climate change has forced us to innovate in the past and is forcing us to do so again. Here lies the next revolution. Thomas Woltz believes that by working with scientists, landscape architects might make a difference to agriculture.[84] This basic desire has been the focus of this chapter because landscape architects are trained to deal with the multiple voices that need to be taken into account. Karen Landman makes the important point that it is hard 'to get young people interested in farming because those raised on a farm know about the economic challenges and those raised in the city don't know where to begin'.[85] The young farmer and designer at the beginning of this chapter grew up on her family farm, but the challenge is for new designers who do not have such a background to begin to work with the many opportunities that await, and to create them, working with multiple, not single voices. It is an uncertain path.[86]

Design georgics is a difficult arena and suggests long-term, rigorous, and complex projects, and new manners of working to those that have come more easily to designers.

Notes

1 *The Fables of Aesop*, The Folio Society, 1998. The fables are credited to Aesop, a slave, in Greece approximately 620–560 BC.
2 This is Christie Stewart. Her parents are to her left, background. Image by Paul Verity; permission P. Verity and Christie Stewart.
3 Richard Weller, 2014, 'Stewardship now? Reflections on landscape architecture's raison d'etre in the 21st century', *Landscape Journal* 33(2): 85–108.
4 For example, in Bach's great cantata BWV 147 *Herz und Mund und Tat und Leben*, the chorus *Wohl mir, dass ich Jesum habe*, where both the main vocal line and the continuo line are equally framed and equally singable – both are equal voices of interest and are not in competition but harmony. Bach's cantatas and 'three part inventions' for piano are noted examples of counterpoint. For those graphically minded, they might like to see the performances and graphical representations of Bach's counterpoint by Stephen Malinowski. The word counterpoint is from the Latin word '*contrapunctus*' means 'point against point'.
5 It will be noted later in this chapter that the imperative of urban agriculture or community gardens competes with biodiversity, or habitat provision for local species.
6 See Catherine Bertini, 2015, 'Invisible women', *Daedalus. Journal of the American Academy of Arts & Sciences* 144(4): 24–30, for a clear account of the role of women in assisting to shift the combined poverty and low nutrition of the rural poor. She notes that 'women eat last', and the impacts that this has had.
7 UN Rural Poverty Report 2011, p. 9. Available online at http://www.ifad.org/rpr2011/report/e/rpr2011.pdf. Accessed 4 December 2015. Of note, and worth quoting in full here, is that 'During the period between September 2006 and June 2008, international food prices almost doubled' (p. 30); the impacts of this are that: 'In many countries, low-income people found themselves unable to properly feed themselves or their children. Across the world, poor households resorted to taking children (often especially girls) out of school, selling their livestock assets, switching to less nutritious, more filling and cheaper food and cutting down on non-food expenses. FAO estimated in 2008 that the price spike had added about 100 million to the global number of hungry people. Those affected were not just in Asia, home to the largest number of hungry people (640 million) or sub-Saharan Africa, where is found the highest prevalence of under-nourishment relative to its population (32 per cent). The largest percentage increases in the number of hungry

people in 2009 relative to 2008 were actually in the Middle East and North Africa (an increase of 14 per cent) and Latin America and the Caribbean (an increase of 13 per cent).'
8   Dewey Thorbeck, 2012, *Rural Design: A New Design Discipline*, London: Routledge, p. xv.
9   Ibid.
10 Graeme Barker and Candice Goucher (eds), 2015, *Cambridge World History, Vol. 2: A World with Agriculture, 12,000 BCE–500 CE*, Cambridge: Cambridge University Press, p. 1.
11 M.E. Kislev, D. Nadel, and I. Carmi, 1992, 'Epipalaeolithic (19,000BP) cereal and fruit diet at Ohalo II, Sea of Galilee', *Review of Palaeobotany and Palynology* 73: 161–166.
12 Odile Peyron, Joël Guiot, Rachid Cheddadi, Pavel Tarasov, Maurice Reille, Jacques-Louis de Beaulieu, Sytze Bottema, and Valérie Andrieu, 1997, 'Climatic reconstruction in Europe for 18,000 YR B.P. from pollen data', *Quaternary Research* 49(2): 183–196.
13 As an example of this, 30,000 years ago new toolkits on the Indian subcontinent were adopted as a consequence of the increasing size of human populations and resource stress due to natural environmental deterioration; see M. Petraglia, C. Clarkson, N. Boivin, M. Haslam, R. Korisettar, G. Chaubey, P. Ditchfield, D. Fuller, H. James, S. Jones, T. Kivisild, J. Koshy, M. Lahr, M. Metspalu, R. Roberts, and L. Arnold, 2009, 'Population increase and environmental deterioration correspond with microlithic innovations in South Asia *c*.35,000 years ago', *Proceedings of the National Academy of Sciences, USA* 106(30): 12261–12266. Other work done in India has supported the same conclusions – that difficulties such as increased aridity 4,000 years ago spurred the development of sedentary agriculture. Camilo Ponton, Liviu Giosan, Tim I. Eglington, Dorian Q. Fuller, Joel E. Johnson, Pushpendra Kumar, and Tim S. Collett, 2012, 'Holocene aridification of India', *Geophysical Research Letters* 39(3), available online at doi:10.1029/2011GL050722, found that sedentary agriculture took hold in the drying central and south India, and the establishment of a more variable hydro-climate may have led to the rapid proliferation of water-conservation technology in southern India.
14 Nick Brooks, 2006, 'Cultural responses to aridity in the Middle Holocene and increased social complexity', *Quaternary International* 151: 29–49.
15 A discussion of the relationship of climate stress on human activity and organisation towards agriculture is found in the details of Renée Hetherington and Robert G.B. Reid, 2010, *The Climate Connection: Climate Change and Modern Human Evolution*, Cambridge: Cambridge University Press, pp. 245–268. See also Hélène Jousse, 2006, 'What is the impact of Holocene climatic changes on human societies? Analysis of West African Neolithic populations dietary customs', *Quaternary International* 151: 63–73: she found that the increasing stress of aridity appeared to be a forcing factor leading to cultural changes and social development. See also Brooks, 2006; he notes that increased social complexity was largely driven by environmental deterioration in Mesopotamia, the Indus-Sarasvati region, northern China, and coastal Peru, but suggests that higher resolution palaeoenvironmental data are required. He considers that the view of adaptation to climate change as a means of neutralising the impacts of environmental change is naïve.
16 The issue of coping with the speed and direction of change that occurred at the end of the Pleistocene is discussed by Clive Finlayson, 2004, *Neanderthals and Modern Humans: An Ecological and Evolutionary Perspective*, Cambridge: Cambridge University Press. Finlayson believes that *Homo sapiens* survived

17 Joy McCorriston and Frank Hole, 1991, 'The ecology of seasonal stress and the origins of agriculture in the Near East', *American Anthropology* 93: 46–69.
18 Alan H. Simmons, 2007, *The Neolithic Revolution in the Near East: Transforming the Human Landscape*, Tuscon, AZ: University of Arizona Press. In chapter 2, pp. 10–29, Simmons gives a full appraisal of theories and ideas about why people became food producers.
19 Daniel Zohary, Maria Hopf, and Ehud Weiss, 2012, *Domestication of Plants in the Old World: The Origin and Spread of Plants in South-West Asia, Europe, and the Mediterranean Basin*, 4th edn, Oxford: Oxford University Press, pp. 1–3.
20 Ibid., p. 2.
21 Thorkild Jacobsen, 1978, *The Treasures of Darkness: A History of Mesopotamian Religion*, New Haven, CT: Yale University Press, p. 10.
22 Charles R. Clement, William M. Denevan, Michael J. Heckenberger, André Braga Junqueira, Eduardo G. Neves, Wenceslau G. Teixeira, and William I. Woods, 2015, 'The domestication of Amazonia before European conquest', *Proceedings of the Royal Society B* 282, available online at doi:10.1098/rspb.2015.0813.
23 Gustavo Santos Vecino, Carlos Albeiro Monsalve Marín, and Luz Victoria Correa Salas, 2015, 'Holocene transition and plant cultivation from the end of early Holocene through middle Holocene in Northwest Colombia', *Quaternary International* 363: 28–42.
24 Manuel Arroyo-Kalin, 2010, 'The Amazonian Formative: Crop domestication and anthropogenic soils', *Diversity* 2: 473–504. This paper discusses anthropogenic landscape transformations in Amazonia and reveals the contested ideas about the emergence of sedentism and agriculture in this region.
25 Francis E. Mayle and Mitchell J. Power, 2008, 'Impact of a drier Early–Mid-Holocene climate upon Amazonian forests', *Philosophical Transactions of the Royal Society B* 363: 1829–1838. This volume is devoted to climate change in the Amazon. The Amazon region changed its vegetative composition markedly all through human occupation.
26 Mauro Ambrosoli, 1997, unfolded these great changes in *The Wild and the Sown: Botany and Agriculture in Western Europe 1350–1850*, Cambridge: Cambridge University Press.
27 Robert E. Evenson and D. Gollan, 2003, 'Assessing the Impact of the Green Revolution, 1960 to 2000', *Science* 300(2 May): 5620.
28 Vandana Shiva, 2016, *The Violence of the Green Revolution: Third World Agriculture, Ecology and Politics*, Lexington, KY: University Press of Kentucky. The focus of Chapter 2, 'Miracle seeds' and the destruction of genetic diversity, highlights lost crop genetic diversity.
29 If available internationally, a lovely documentary is *The Seed Hunter*, 2008, Australian Broadcasting Commission. This tracks a Syrian, Australian, US, Russian, and Tajik team on the search for wild cultivars, notably chickpeas, in the mountains of Tajikistan.
30 Molly E. Brown and Chris C. Funk, 2008, 'Food security under climate change', *Science* 319(5863): 580–581.
31 John Holmes, Bernie Jones, and Brian Heap, 2015, 'Smart villages', *Science* 10.1126/science.aad6521.
32 Available online at http://e4sv.org/about-us/what-are-smart-villages/ Accessed 4 March 2016. That gender equality and democracy are given suggests that the ambitions of the organisation lie beyond food.

(continued from previous note) because we became increasingly inventive (pp. 203–208), and shortages of herbivores and other stresses led to agriculture (p. 203).

33 As was noted for the United States by Wendell Berry, 1977, *The Unsettling of America: Culture & Agriculture*, San Francisco, CA: Sierra Club. Berry argued that the large-scale agribusinesses were taking farming out of its cultural context and away from the families which have driven agriculture and agricultural innovation, including care of the land. It is important to keep farming families on the land as a spiritual discipline.
34 Available online at http://e4sv.org/jaideep-prabhu-cambridge-judge-business-school-discusses-the-frugal-innovation-revolution-that-is-taking-the-world-by-storm/ Accessed 4 March 2016.
35 Dewey Thorbeck, 2012, discusses this in relation to the USA, in *Rural Design: A New Discipline*, London: Routledge, pp. 205–222.
36 Some research showed that potatoes could grow on simulated Mars soil, but only for 50 days. G.W. Wieger Wamelink, Joep Y. Frissel, Wilfred H.J. Krijnen, M. Rinie Verwoert, and Paul W. Goedhart, 2014, 'Can plants grow on Mars and the Moon: A growth experiment on Mars and Moon soil simulants', *PLOS-One*: doi:10.1371/journal.pone.0103138. Another consideration is that using human excreta directly on plants is dangerous to human health due to the pathogens contained in our faeces.
37 W.H. van der Putten, 2005, 'Plant-soil feedback and soil biodiversity affect the composition of plant communities', in R.D. Bardgett, M.B. Usher and D.W. Hopkins (eds), *Biological Diversity and Function in Soils*, Cambridge: Cambridge University Press, pp. 250–272.
38 A. Sugden, R. Stone, and C. Ash, 2004, 'Ecology in the underworld', *Science* 304: 1613.
39 Keith Paustian, Johannes Lehmann, Stephen Ogle, David Reay, G. Philip Robertson and Pete Smith, 2016, 'Climate-smart soils', *Nature* 532: 49–57. The authors note re Green House Gas emissions (GHG) that: 'Targeted basic research on soil processes, expanding measurement and monitoring networks, and further developing global geospatial soils data can improve predictive models and reduce uncertainties. Ongoing advances in information technology and complex system and "Big Data" integration offer the potential to engage a broad-range of stakeholders, including land managers, to "crowd-source" local knowledge of agricultural management practices through web-based computer and mobile apps, and help drive advanced model-based GHG metrics. This will facilitate the implementation of climate-smart soil management policies, via cap-and-trade systems, product supply-chain initiatives for "low-carbon" consumer products, and national and international GHG mitigation policies; it will also promote more sustainable and climate-resilient agricultural systems, globally.' I address the potential of designers to engage with these developments later in the chapter.
40 David F.L.S. Norton and Nick Reid, 2013, *Nature and Farming: Sustaining Native Biodiversity in Agricultural Lands*, Collingwood, Australia: CSIRO Publishing, p. 17.
41 Ibid., p. 33.
42 J. Esquinas-Alcázar, 2005, 'Protecting crop genetic diversity for food security: Political, ethical and technical challenges', *Nature Reviews. Genetics* 6: 946–953.
43 For a general discussion on the state of world crops see: http://www.idrc.ca/EN/Resources/Publications/Pages/ArticleDetails.aspx?PublicationID=565, accessed 2 October 2015. They note that 80 per cent of Africa's food comes from such small-scale farmers. The majority of these farmers in poor regions are women (sometimes 80 per cent). For a discussion of women in this situation, and suggested avenues of assistance, see: Catherin Bertini, 2015, 'Invisible women', *Daedalus. Journal of the American Academy of Arts & Sciences*:

24–30. This edition was dedicated to The Future of Food, Health & the Environment of a Full Earth.
44 Ibid. This data is from the International Development Research Centre, Canada.
45 Landline Australian Broadcasting Corporation, 2015, *The Future of Food*, http://www.abc.net.au/landline/content/2010/s3253782.htm Accessed 24 November 2016.
46 Jeffrey L. Bennetzen, 2013, 'Crop diversification for improved agricultural resilience during rapid climate change', *Plant and Animal Genome XXI Conference 2013*.
47 For an excellent review of the potentials of old crops somewhat neglected and their current status see: Festo Massawe, Sean Mayes, and Acga Cheng, 2016, 'Crop diversity: An unexploited treasure trove for food security', *Trends in Plant Science* 27(5): 365–368. In the paper the authors note the major under-utilised crops of buckwheat, pearl millet, quinoa, kiwicha (all cereals), jackfruit, mangosteen, Chinese leek, caigua (fruit and vegetables), mashua tuber, rootstock, taro, yam (tubers and roots), and Bambara groundnut, petai, jering and winged bean (pulses). They point out that some of these under-utilised crops with good nutrition have unique cultural and crop traits that make them invaluable to local farmers.
48 Michael Abberton, 2016, 'Climate change and orphan crops in West Africa', *Plant and Animal Genome XXIII Conference*; this work is available as a PowerPoint; the work is being done at the International Institute of Tropical Agriculture (IITA) in Ibadan, Nigeria. Abberton includes information about soil degradation and child malnutrition, a great impetus for the use of high protein locally indigenous crops to be resurrected or to have the same intense research and manipulation that our main, originally wild crops (such as maize) have had.
49 From the transcript of the Australian Broadcasting Commission movie *The Seed Collector*, which follows agronomist Dr Ken Street to Tajikistan. See also: S. Ceccarelli, 2012, 'Landraces: Importance and use in breeding and environmentally friendly agronomic systems', in N. Maxted, *et al.* (eds), *Agrobiodiversity Conservation: Securing the Diversity of Crop Wild Relatives and Landraces*, Wallingford, UK: CAB International, pp. 103–117.
50 Of course, some regions would have little topographic variation, notably in the great cereal-growing regions of the world.
51 Lenore Fahrig, Jacques Baudry, Lluís Brotons, Françoise G. Burel, Thomas O. Crist, Robert J. Fuller, Clelia Sirami, Gavin M. Siriwardena, Jean-Louis Martin, 2011, 'Functional landscape heterogeneity and animal biodiversity in agricultural landscapes', *Ecology Letters* 14: 101–112.
52 For discussion of finer-grained heterogeneity and refugia on the farm see Laura Henckel, Luca Börger, Helmut Meiss, Sabrina Gaba, and Vincent Bretagnolle, 2015, 'Organic fields sustain weed metacommunity dynamics in farmland landscapes', *Proceedings of the Royal Society B* 282, available online at doi:10.1098/rspb.2015.0002.
53 Donald Worster, 2003, 'Watershed democracy: Recovering the lost vision of John Wesley Powell', *Journal of Land Resources & Environmental Law* 23: 57–66.
54 Elizabeth Barham, 2001, 'Ecological boundaries as community boundaries: The politics of watersheds', *Society and Natural Resources* 14: 181–191.
55 In Chapter 2 I discussed the ideas of holdouts and micro-refugia to deal with future climatic challenges. These ideas can be taken into agricultural landscapes.
56 Tara G. Martin and James E.M. Watson, 2016, 'Intact ecosystems provide the best defence against climate change', *Nature Climate Change* 6: 122–124. The authors point out that 'Humans are adapting to climate change, but often in

ways that further compound our effects on nature, and in turn the impact of climate change on us'.
57   A full discussion of land-sharing and land-sparing is given in: Andrew Balmford, Rhys Green and Ben Phalan, 2015, 'Land for food & land for nature?', *Daedalus* 144(4): 57–75.
58   Some cities are also mapping the potential for accommodating climate change despite the constraints of built form, which is a central problem of any city and suburban region. Cape Town, rightly proud of sitting amid the world's most profound global biodiversity landscapes, the Cape Floristic Region, has used conservation methodologies to define the best configurations for a Biodiversity Network (BioNet), and research is being done to articulate the spatial positioning for biodiversity to be maintained. Urban development and invasive alien species are the major threats to regional biodiversity in the Cape Floristic Region, where 70 per cent of the 9,600 species are unique to the region.
59   Herman A. Karl, Lynn Scarlett, Juan Carlos Vargas-Moreno, and Michael Flaxman, 2012, 'Synthesis: Developing the institutions to coordinate science, politics, and communities for action to restore and sustain lands', in Herman Karl, Lynn Scarlett, Juan Carlos Vargas-Moreno, Michael Flaxman (eds), *Restoring Lands – Coordinating Science, Politics and Action: Complexities of Climate and Governance*, available online at doi:10.1007/978-94-007-2549-2_22, Springer Science+Business Media B.V., p. 498.
60   See http://dirt.asla.org/2011/11/01/restoration-ecology-in-agrarian-landscapes/ Accessed 10 March 2016.
61   Ibid.
62   An excellent summary of the issues involved – it is not just crops – is given in: Patrick Meyfroidt, Florian Schierhorn, Alexander V. Prishchepov, Daniel Müller, and Tobias Kuemmerle, 2016, 'Drivers, constraints and trade-offs associated with recultivating abandoned cropland in Russia, Ukraine and Kazakhstan', *Global Environmental Change* 37: 1–15.
63   Megan C. Evans, Josie Carwardine, Rod J. Fensham, Don W. Butler, Kerrie A. Wilson, Hugh P. Possingham, and Tara G. Martin, 2015, 'Carbon farming via assisted natural regeneration as a cost-effective mechanism for restoring biodiversity in agricultural landscapes', *ScienceDirect* 50: 114–129.
64   From the conference *Soil, Big Data and the Future of Agriculture*, United States Study Centre of the University of Sydney, Canberra, June 2015.
65   A distinction must be made between using data to redesign and the traditional farm adviser; designers have no training for advice about farm production methods because it is a specialised focus within agricultural training.
66   The use of data and digital technologies has recently been discussed by Jillian Walliss and Heike Rahmann, 2016, *Landscape Architecture and Digital Technologies: Re-conceptualising Design and Making*, Abingdon: Routledge.
67   IBM Research Brazil note, available online at http://www.research.ibm.com/articles/precision_agriculture.shtml; accessed 5 October 2015.
68   Comments from Jolene Otway, *Soil, Big Data and the Future of Agriculture Conference 2015* in Canberra, The University of Western Australia Institute of Agriculture, August 2015.
69   Jane Amidon, 2012, 'Two shifts and four threads: Economic and ecologic challenges for landscape architecture and urbanism', *Topos: European Landscape Magazine* 80: 16–24.
70   *Oxford Latin Dictionary*, Oxford University Press.
71   Laura Lengnick, 2015, *Resilient Agriculture: Cultivating Food Systems for a Changing Climate*, Gabriola Island, BC: New Society Publishers. This book addresses sustainable agriculture and works with how we join the various

72 Marianne E. Krasny and Keith J. Tidball, 2015, *Civic Ecology: Adaptation and Transformation from the Ground Up*, Cambridge, MA: The MIT Press.

73 Thomas A. Lyson, 2012, 'Civic agriculture', in Raymond De Young and Thomas Princen (eds), *The Localization Reader: Adapting to the Coming Downshift*, Cambridge, MA: The MIT Press, pp. 117–128.

74 Karen Landman, 2013, 'Local food spaces: Constructing communities of food', *Local Environment* 18(5): 521–641. Available online at https://www.youtube.com/watch?v=XFj_ip0mM04. Accessed 30 January 2016.

75 Cited in Bridget Keane, 2015, 'Parallel genealogies: Chris Reed', *Landscape Architecture Australia* 148: 23.

76 Some work with depressed social conditions can be seen in Nelson Byrd Woltz shown in: Warren T. Byrd and Thomas L. Woltz, 2013, *Nelson Byrd Woltz: Garden, Park, Community, Farm*, New York: Princeton Architectural Press; see also the many organisations working with community gardens.

77 Apparently noted by an early twentieth-century British ambassador to the US.

78 Discussed by Nathan Stephenson, 2014, 'Making the transition to the third era of natural resources management', *George Wright Forum* 31: 227–235. A period of intense change of ambitions and management is upon us in national parks. Stephenson refers to this as the third era of management for national parks.

79 Robert B. Keiter, 2013, *To Conserve Unimpaired: The Evolution of the National Park Idea*, Washington, D.C.: Island Press/Center for Resource Economics.

80 This shift has been described and discussed on a quite personal level by Stephenson, 2014. This work was for the National Park Service Centennial 2016.

81 Richard West Sellars, 1997, *Preserving Nature in the National Parks: A History*, New Haven, CT: Yale University Press.

82 I discuss this tendency in: Grose, 2014, 'Gaps and futures in working between ecology and design for constructed ecologies', *Landscape and Urban Planning* 132: 69–78.

83 For example, foci on transport have driven ideas about the compact city and densification all over the world as if a cure-all to suburban sprawl; yet in many cities densification has led to infill housing and the loss of large trees that sequester carbon – with decreased biodiversity, increased impermeable surfaces and subsequent increased water run-off, and a general loss of green spaces per head of population, despite population commonly being the basis on which public open spaces were originally calculated, not percentage land area. I discuss this in relation to Western Australia in Grose, 2007, 'Perth's Stephenson-Hepburn plan of 1955, 10% POS, and housing then and now', *Australian Planner* 44(4): 20.

84 Dakotah Bertsch, 2011, 'Restoration ecology in agrarian landscapes', interview reported in http://dirt.asla.org/2011/11/01/restoration-ecology-in-agrarian-landscapes/ Accessed 10 January 2016.

85 Teresa Pitman, 2009, 'Landscape architecture professor travels 18,000 kilometres across the North America to study urban agriculture', *City Farmer News*. Available online at http://www.cityfarmer.info/2009/11/14/landscape-architecture-professor-travels-18000-kilometres-across-the-north-america-to-study-urban-agriculture/ Accessed 16 March 2016.

86 The uncertain paths for national parks is the topic of William C. Tweed's 2010 book *Uncertain Paths: A Search for the Future of National Parks*, Berkeley, CA: University of California Press.

5

INQUIRIES, NOT ASSUMPTIONS

Every day you play with the light of the Universe.
Pablo Neruda (1924), 'Every day you play'[1]

Long ago in France, someone began to paint in the flickering darkness of a cave lit by charcoal or fat.[2] Their subject was of immense importance to them because it concerned the vital connections of the seasons, and the signs in the stars told of the availability of meat, fat, hides, horns and hoofs, and protein.[3] The painters had noted that the cluster of stars called the Pleiades were linked to the seasonal changes of the great wild aurochs.[4] The image left to us on the cave wall is from a world with little artificial light when the stars dominated the night view.

The tradition that the Pleiades were bright female stars recurs across the globe. A number of mythologies refer to the Pleiades – the Old Testament of the Bible;[5] both the *Iliad* and the *Odyssey*;[6] legends of Aboriginal Australians and peoples in Oceania;[7] and texts in China, North and South America, the Middle East, and Africa. In very recent times, the North Star, or Polaris, was the lodestar in the Northern Hemisphere that led to great voyages of discovery across the world.[8] Polaris drops out of sight as one crosses the Equator and in the Southern Hemisphere, a completely different sky, no single brightest star at the poles guided travellers.[9]

Although comets such as Hale–Bopp still amaze, and the Aurora Borealis and Aurora Australis incite tourism and wonder,[10] our view of the night sky is now so vastly reduced by artificial light at night that in the developed world we have few opportunities to look at the stars much at all. Yet our galaxy, the Milky Way, is a gravitationally bound collection of roughly 100 billion stars, seen most clearly in the Southern Hemisphere.

In the last 200 years the change to extensive night lighting has been rapid and has been one of the greatest changes in human life. Discussion of light and lighting in design schools, if considered, is almost entirely at the scale of lighting fixtures, the creation of light-scapes of colour and supposed aesthetics, or of present and imagined energy savings. In other disciplines, and in 'sustainability' research, public lighting is confined to energy, with its associated carbon-emission reduction[11] and monetary savings, lighting provision, and grids. However, major issues beyond energy have arisen. Important new evidence from sleep science, physiology, ecology, crime and road safety, networked sensor systems, lighting technologies and physics, and astronomy will transform our decisions about public lighting. The extraordinary range of impacts of light at night on human health (and wildlife) are set to change our views of the meaning of artificial light at night[12] and will propel designers towards a redesign of streets. Long-held assumptions – that light is good or benign, that more light is better, that light prevents crime, that 'the public' wants more light, and that darkness is bad – need to be cast aside. All of these are incorrect or unfounded.

There is an urgency[13] to reassess old assumptions. Throughout the world, with little or no design, new public lights are replacing old ones. There is currently a rapid retrofitting to LED (light emitting diode) with the sole focus on energy because it is assumed to be the only factor of importance. New information has not reached those who need it, with decisions continuing to use old assumptions about light at night. In this chapter I outline the impacts of artificial light at night on human health and ecological processes. Some impacts will

INQUIRIES, NOT ASSUMPTIONS

present global challenges to communities and will inform critical revisions in health and urbanism of vital importance to landscape architecture and urban design.

Playing with the light of the universe

Since the invention of fire, we have illuminated darkness. Artificial light has been generated by dung, wood, oil, dried fish, beeswax, peat, coal, and gas.[14] With the invention of the electric light, we have been playing very seriously with the light of the universe. We can see the lit world of today everywhere. Much of it is wasted light, as seen clearly in Figure 5.1.

We have changed the night, and perhaps even the nature of darkness and what it means to us. Darkness can be defined; a dark

Figure 5.1 A now common and typical view of Earth: the Atlantic coast of the US from the Space Station in 2013. Note the Great Lakes, the cities of New York, Baltimore, and Philadelphia, as well as others more distant. Accessed 6 June 2016. All of this upward light is wasted light. Upward light in the USA alone costs approximately $3.3 billion per year, all for the benefit, it would seem, of the handful of people in the International Space Station. This image is therefore an extremely expensive one to achieve, and needs to be seen as a sign of extraordinary waste.[15]

Image: Courtesy of NASA/JPL-Caltech, 2012, 'Eastern coast of United States at night'. http://www.nasa.gov/mission_pages/station/multimedia/gallery/iss030e078095.html.

bedroom is one where you cannot see the other side of the room. It is getting harder to find darkness, and few people who live in cities can regularly find darkness outside. Many might no longer believe that you can see your shadow on the ground from the Moon, or have forgotten that starlight is in our vocabulary for a reason. With the proliferation of the electric light, permanent night-time illumination has become commonplace.

At present, our Sun is the only star that we can see in major cities, such as New York and London. Yet in dark natural conditions about 2,700 stars should be visible with the naked eye in the Northern Hemisphere; more in the Southern. My city is typical of the developed world; in Melbourne's middle suburbs, only about 2–3 per cent of the stars are visible now, whereas in the 1880s all would have been visible – that is, in just 140 years, we have lost from our vision 97–98 per cent of our stars. Loss of visibility of the night sky is increasing exponentially at the same rate as the growth in outdoor illumination, and it is predicted that in a few generations very few people will be able to see any stars at all.[16] Yet many believe that in not looking at the stars much at all, and not being able to see them, we forget who we are. How we are lighting our planet, and how we might light it in the future are moral and philosophical questions now being considered around the world.

No one has ever debated the question of whether we want to lose our view of the night sky. No referendum has ever been held on this question. The change that we are experiencing reveals the cumulative effects of thousands of small decisions – of, for example, the folks who put a spotlight on outside and leave it on most nights – that together impose a profound change in the human view of the night sky.[17]

How is light pollution?

Light pollution is a term for any unwanted or nuisance or trespass artificial light that has adverse effects. The general expression, 'light pollution', means the alteration of natural light levels in the night environment due to the introduction of artificial light.[18] Astronomical light pollution, when light spills into the night sky, impacts the visibility of the stars. Even if we shield lights to avoid astronomical light pollution, ecological light pollution can still occur because light alters and impacts ecological processes and animal behaviour in the local area. More than ten years ago, Travis Longcore and Caroline Rich published *Ecological Consequences of Artificial Light at Night*, and alerted the world to this new research arena.[19] Designers have yet to take heed.

The brightening of the night sky caused by all lights with any upward component, either directly or indirectly, is called skyglow.[20] Street lighting, vehicle lights, traffic signals and signs, shops, architectural floodlighting, illuminated billboards, internal lights escaping into the night through open curtains, porch lights left on, and careless use of private and business spotlights all contribute to skyglow.

The spectrum (colour) of light matters

When we go to the supermarket and buy new globes for the light fittings in our homes, we find a wide choice, as in yoghurt, chocolate, or cheese. But while food is a matter of our own taste or finances, most of us are relatively ignorant of the notes on the sides of light-globe packets. What are the differences in warm white and cool white? The letter K refers to Kelvin, a numerical expression of where light produced by the light-globe occurs on the colour spectrum (Figure 5.2). The mathematical physicist and engineer Sir William Thomson (Lord Kelvin) established the relationship of colour to temperature when he burnt a block of carbon and recorded its colour as it heated. He found that the colour of carbon changed with temperature and that the higher the temperature, the bluer the carbon became. Colour temperature also gives a different look to the light emitted. High Kelvin, in the blue part of the spectrum, appear white-blue; cool white has a higher colour temperature than warm white, just as a white flame is hotter than a red. When globes of a high K are selected for streetlights, they give a cold, often harsh light.

Many designers appear unaware of colour temperature. Recently, a friend horrified me because the architect of her home renovation had put in a row of 7,000K LED lights in her kitchen. As you can see in the diagram below, 7,000K is far into the blue range of light. Did the architect think my friend was doing surgery on her kitchen table? Blue light is bad for human health at night because it impacts our melatonin production. Melatonin, discussed later in this chapter, is a natural neuro-hormone that is released by darkness and inhibited by light. It is a major player in how light impacts animal life and a warning barometer for human health. While any sort of light impacts melatonin, the blue portion of the spectrum is most potent in melatonin disruption. Medical science now shows very clearly that we should avoid blue light at night. Many lights which appear white to the eye emit large amounts of blue light. For example, '29% of the spectrum of 4,000K LED lighting is emitted as blue light, which the human eye perceives as a harsh white colour'.[21]

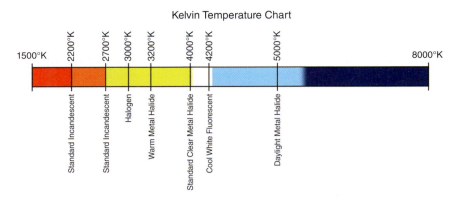

Figure 5.2 A Kelvin colour temperature chart, showing some standard light types.
Source: http://darksky.org/the-promise-and-challenges-of-led-lighting-a-practical-guide/.

Ecological aspects of the effects of artificial night lighting

Observers have noted the ecological effects of light at night on animals for at least 200 years, since lighthouse keepers recorded that birds struck lanterns more frequently when clouds did not permit a view of the stars.[22] The height and intensity of city lights and horizon glows are important distractions for migrating birds.[23] Less well known are the effects of artificial night lighting on the mating flights of ants, disruption of zooplankton, altered nesting patterns of birds, and changed movement of mammals, particularly predators, and disturbance to fish foraging and predation.[24]

Artificial light at night and street-trees

Plants have phototropism – that is, their shoots grow towards the light, and roots have geotropism, growing towards the earth, away from light. Despite the role of trees as a central component of the street scene, the effects of artificial night lighting on plants remain almost unexamined. A pioneering study as long ago as 1936 in New York[25] noted that electric lights altered leaf retention. Carolina poplar (*Populus canadensis*), London plane (*Platanus acerifolia*), sycamore (*Platanus occidentalis*), and crack willow (*Salix fragilis*) – close to 76-watt,[26] 11-volt electric street lights – retained leaves with the onset of winter in that portion of the tree that was artificially illuminated.[27] Since then, a number of studies on trees have also shown that trees grown near streetlights have a prolonged photoperiod and grow for longer (and keep their leaves for longer if deciduous) than plants not exposed to streetlights.[28] Recent work in the UK has found that light pollution produces earlier tree budburst.[29]

Light is central to a great range of plant function and physiology.[30] Plants, like humans, have a circadian system, or circadian rhythm.[31] In plants, this 'clock' enhances resistance to insects and pathogens and controls the levels of a range of important substances found in the leaves that we eat.[32] While we might be amused to discover that cabbage has a circadian system like us,[33] our commonality with cabbage shows us that melatonin evolved very early with life on earth and is fundamental and ubiquitous to all life. It is therefore not surprising that disruption of plant growth occurs with artificial light.[34,35] Research needs to be done on whether artificial light at night is an added stress for trees as they cope with climate change.

New knowledge about animals and artificial light at night

As insects are fundamental basal members of food chains, any negative impacts upon them are important for entire ecosystems. We know moths are fatally attracted to light. Artificial lights attract and kill millions of insects every night in Germany,[36] with the number of insects killed over a single summer being in the order of 10 billion. Though this level of death would have an ecological impact on the insects, their predators, and pollination – ecological impacts remain largely absent from discussion of lighting provision or lighting standards in cities and suburbs (and thus sits at odds with biodiversity objectives). Sensitive to light via their circadian system, insects are also affected by the spectrum of the light. The 'wrong' colour of a streetlight has strong impacts on biodiversity, cascading impacts throughout entire ecosystems.[37]

Lights thus carry ecological penalties. Designers often fail to realise the ecological implications of the light they propose. With no heed to the ecological impacts of artificial light, it is common to see strong lights for structures, novel events (such as the lighting memorial to the Twin Towers in New York),[38] and public shows, up-lighting on buildings, and offices in high rises where there are no blinds. This disregard is playing with the light of the Universe.

The spectral composition of light is important in a range of physiological mechanisms in animals, and current shifts towards whiter public lighting are likely to increase negative environmental impacts.[39] A simple and recent study serves to show this. Garden Island is a little sanddune and limestone island off the Western Australian coast. Used as a navigational point by the Dutch East India Company's ships in the seventeenth century, it became a place of holiday shacks in the twentieth. Then, in 1978, its public use was restricted when it became the Australian Navy's west coast frigate and submarine base, Australia's largest naval base. The island is well wooded and remains the home of rare tammar wallabies, animals that look like miniature kangaroos. Research on tammars by Australian and German zoologists shows a startling shift in the reproductive pattern of wallabies living near the highly lit naval base from those living in adjoining dark bushland.[40] For the first time, clear evidence indicates that artificial light at night desynchronises seasonal physiological processes.[41] In this important paper, the authors noted that the replacement of high-pressure sodium lights with the strong white light of LED will suppress melatonin to a greater extent and therefore impact biodiversity. Alerts about the strong impacts of artificial light at night on invertebrates and vertebrates edges us towards threats to the human species; as we become more aware, we should begin to change street design and the design of the night landscape.

The human mammal

For decades, scientists studied bats, moths, turtles, and birds in regard to disruption and death by artificial lighting. If zoologists note the impacts of artificial light on fish, we need to consider the effects of night lighting on *fishermen*. There is now mounting evidence in the scientific medical community that artificial night lighting levels have strong impacts on humans and is likely a sleeping giant threatening human health. Night lighting can interrupt the human circadian system and consequently impacts our physiology, metabolism, and behaviour. I have found that this knowledge comes as a surprise to most people because few have thought about light as potentially bad, or that it needs careful control. Inside our homes and workplaces we light excessively; in the developed world the light levels commonly used inside the home are enough to disrupt our circadian system to such a degree that we can be considered to be permanently suffering mild jet lag.[42]

New knowledge about light and physiology: Getting out of our rhythm

Melatonin drives the circadian system that regulates sleep, mood, and satiety. Melatonin tells the body that it is dark and sets the human 'body clock'; human melatonin levels have evolved to be produced in darkness between 21.00hr and 07.00hr but melatonin is largely absent when the body is exposed to light during the day. The rate of production of melatonin increases after midnight.

Light exposure at night, especially in the 'small hours', can flat-line melatonin production. As little as 0.4 photopic lux of blue light can reduce melatonin in humans.[43] In ordinary domestic situations, light exposures are often white rather than blue. But exposures of less than

100 lux of white light can also have an appreciable effect on melatonin. Sometimes we are exposed to very high levels of light and might not realise it; Barry Clark, former Government Astronomer in Victoria, mentions that evening exposure to 500 to 1,000 lux for one to two hours is typical for sporting teams and spectators in high-level competition and televised games at night.[44]

New knowledge about the workings of the eye

We have all heard about rods and cones as two types of light receivers in the eye and have assumed that they defined sight as the sole function of the eye. For many decades, knowledge of rods and cones helped to develop standards for public lighting and the lighting requirements of homes and offices. However, in 2002, researchers found a new type of photoreceptor in mice, and then in humans in 2005.[45] This third type of photoreceptor has revolutionised our understanding of the eye. Like the ear, which has the two functions of hearing and balance, light into the eye is now understood to have two functions. The first function of the eye is for vision (via the rods and cones), and the second is a physiological, circadian function determined by the light received by the eye (via ganglion photoreceptors).

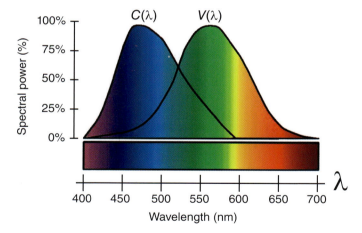

Figure 5.3 This graph shows two simple response curves. The one on the left shows human response to light for our circadian system – showing where we are most responsive – and the one on the right the more commonly seen visual response to the colour of light. It is the circadian system (melanopsin action spectrum, C) that is more important for human health.

Source: From M. Andersen, J. Mardaljevic, and S.W. Lockley, 2012, 'A framework for predicting the non-visual effects of daylight – Part 1: photobiology-based model', *Lighting Research and Technology* 44: 37–53.

THINKING ABOUT DESIGN

The ganglion cell layers are maximally sensitive to blue light – that is, humans are exquisitely sensitive to blue light.[46] This striking new understanding will play a major role in reworking lighting design.[47] As Figure 5.3 suggests, this shift from traditional, vision-based ideas to circadian-based ideas about lighting must ultimately change the design of all next generation guidelines for urban areas, and those for inside buildings,[48] decorative lighting and advertising, and digital signboards.[49] We need to discard old assumptions.

Figure 5.3 shows the significant place of blue light. Yet many new lights are blue-rich, an issue overlooked in the rush to apply only the assumptions of energy efficiency on lighting standards. Blue light can be readily measured, such as the spectrum typical in Australia given in Figure 5.4. A brand new LED light near my house has significantly more blueness, brightness, and glare than any previous light types that had been on the street for more than 100 years.

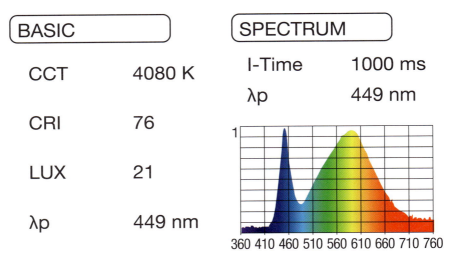

Figure 5.4 Colour spectrum of a new 4,000K LED light in a suburban street, Brunswick, Melbourne, in October 2015. This shows a very high reading of blue light in the 449nm[50] range. It is blue light that most interferes with mammalian circadian rhythms. Blue light keeps us alert during the day; we do not want this at night.[51] This light was very unpleasant to look towards as one walked down the street; the lamp was not flat but in a dropped box, sending more light sideways than other LEDs available. Note also that 4,000K is variable; this one gives a CCT (correlated colour temperature) reading of 4,080K. The CRI is the colour-rendering index that describes how a light source makes the colour appear to the human eye; the higher the CRI the 'better' the colour rendering; above 80 is currently desirable by the lighting industry.

Image: The author.

New knowledge about the health impacts of altered melatonin

Lighting standards often fail our own sense of what our body tolerates and what provides good rest. A few years ago I had a horrid backyard neighbour who left a spotlight on all night and seemed to direct it purposely into my bedrooms. Such issues are common problems between neighbours[52] because quite small amounts of light at night can disturb our sleep when light acts as a trespasser. We often forget that sleep is one of the three pillars of human health, with diet and exercise. Sleep loss and sleep disturbance degrade wellbeing.[53] Daytime sleepiness and fatigue can substantially increase the risk of road and industrial work accidents.[54] In addition, there is evidence that artificial light at night causes 'eveningness' in young people – that is, the circadian clock is delayed later into the night.[55]

We need to be mindful of how light will intrude into sleeping spaces, whether high-rise flats in Hong Kong, sprawling suburbs, or rural hamlets. We need to watch for the type of light and its position, the angle of light, the spectrum, and the height of intrusion into bedrooms (when light is only wanted on the footpath). Some might argue that people can simply close their blinds or curtains if they wish to sleep, but the use of heavy drapes or blackout blinds to block excessive artificial light at night is undesirable because such barriers also block the important beneficial waking effect of fresh morning light,[56] and hinder air circulation.

Disruption of melatonin has wide impacts. We can think of melatonin as keeping our circadian master clock that controls physiological and behavioural processes. Light directly influences mood disorders and cognitive deficits, but until recently we understood little about the finer mechanisms through which light affects these physiological functions.[57] Alterations to sleep due to artificial light at night bring additional physiological impacts that might surprise readers. For example, obesity is associated with increased light at night[58] – sleep curtailment elevates levels of the hunger hormone ghrelin in the human body and acts to stimulate appetite.[59]

The greatest concern with artificial light at night lies with cancer. Breast cancer has risen dramatically in the Western world throughout the twentieth century and is the major cause of cancer death in women worldwide. Thirty years ago the increase in breast cancer was thought linked with diet, but this has never been substantiated. Some cancers do have a link with some specific trigger: lung cancer with smoking, cervical cancer with human papillomavirus, skin cancer with ultraviolet sun exposure. Major public campaigns have tried to reduce the incidence by dealing with the cause. Breast cancer remains confounding.

This puzzle has led many researchers to ask if there is a major overlooked factor that has changed in the world in the last several decades. The finger is now being pointed at light at night. A leading researcher in the field, Richard Stevens, wrote about the growing evidence linking artificial light at night and cancer in: 'Electric light causes cancer? Surely you're joking, Mr Stevens?'[60] The title plays with the problem of how old assumptions work to disallow new evidence.[61] The idea that light might not be good for us suggests a complete rethink of the night landscape in public areas and in the home – in our laptops, our mobile phones,[62] our televisions, and even that digital clock by the bed. The driver of the apparent link between cancer and artificial light at night is, once again, melatonin.

Melatonin is one of the body's most powerful agents for retarding the growth of breast cancer and other cancers. Many laboratory and field studies have shown reliable positive connections between breast cancer incidence and light exposure at night[63] that led to diminished melatonin. Nightshift nurses are particularly prone, especially if they work in emergency or the operating theatre under intense white lighting. In Israel there is a positive association between night light intensity and cancer rates; high cancer rates occur where intense lighting is used for security, infrastructure, and around transport nodes.[64] Circadian disruption linked with breast cancer is now established.[65] The growing information about the links between melatonin, our circadian system, and cancer now jolt us from any assumption that all light is good or benign. This is an important stage for a major design rethink. The extent and types of artificial light at night currently found in urban design is what the American Medical Association has called 'a man-made self-experiment'.[66] This is strong language indeed and a major whistle-blow for landscape architects, interior designers, architects, urban planners, and designers. We can still have light without the self-experiment.

Questioning assumptions about risk and the meaning of safety at night

We cannot discuss light at night without addressing the issue of safety, but I have done this last not because it is least important, but to place it in terms of a new approach to risk. We need to move thinking about public lighting out of the realm of 'public safety' and into the realm of 'community health', because this is where artificial light's greatest impacts lie. Crime is a public safety issue, but light is a public health issue that the built environment professions have neglected despite a warning more than ten years ago.[67,68] Knowledge about melatonin and other impacts of human health is set to change our concept of risk at night and move it away from crime, accidents, and pedestrian safety to include the association of artificial light at night with sleep disruption, obesity, psychological disturbance, and increased prostate and breast cancer.[69] Such knowledge will move ideas about risk from the short-term issues (small-scale crime, for example) to longer-term ones (life-threatening or debilitating illness), and will require us to reconsider the meaning of risk. More profoundly, it should move discussion about lighting from focusing on energy to working in the realm of urban design for public health.

Despite new knowledge, common belief in the efficacy of lighting against actual crime has remained with the public, and challenging this assumption produces emotive and contentious responses. That light reduces crime is often an opening presumptive statement in publications about urban design. This myth will be enormously hard to shift without good information and an understanding by both the public and urban designers of the difference between real crime and the perceived role of lighting in reducing the *fear of* crime. The difference between these is substantial and can lead to costly errors. The UK government has spent £300 million per year on upgrading street lighting to reduce crime.[70] However, with better lighting, street crime increased in the UK since the programme began.[71] Despite mounting evidence to the contrary, the assumption that more lighting reduces crime persists; councils assume that the public have this assumption, which councils feel compelled to address.

Bob Mizon, in his 2012 book *Light Pollution: Responses and Remedies*, discusses several examples of how external lighting does not deter criminals. Lighting helps the criminal to see what they are doing, to 'case the joint', to determine the risk to the criminal, and to determine home occupancy. As noted by Mizon, insurance companies do not rank external security lights as part of security, because they do not consider that lights deter crime.[72] The claim that lighting does assist crime deterrence by natural surveillance is commonly made, but most writers omit the assistance it simultaneously provides in the commission of crime. By the early 1990s, scientific reviews in the USA and the UK had established that the use of outdoor lighting was ineffectual for actual crime prevention. Because there is growing evidence to support the idea that lighting actually provides net assistance to criminals,[73] it is ironic that people feel safer in lighting conditions that are now known to increase actual crime.[74] At present, there seems to be no way to circumnavigate this confusion based on assumptions.

Light causes glare – the excessive contrast between bright and dark areas in the field of view.[75] Glare prevents criminals being seen because when a light is in our eyes we cannot see. Glare is particularly insidious with LED lights, because many are very bright. Glare will become more important with ageing populations and increased use of LED in public lighting because bright lights blind elderly people. As a personal example, a neighbour behind me in my current home has two LED spotlights that are on motion controls, presumably to pick up a robber. But the lights go on and off all night in response to passing bats, cats, owls, and the orange tree blowing in even a little zephyr of a breeze. I have a view into this property, and if the light comes on, I do not look because the LED lights are too glary for me to see. However, if I saw torchlights in my neighbour's dark yard, I would call the police. This is a typical story of how lights do little to protect and can block the view of a site (and criminals) by neighbours.

Assumptions about energy and lights at night

Lighting accounts for a considerable proportion of the total energy consumption of countries and towns. Although there has been technological improvement in the efficiency of lighting systems, the expansion of street lighting in Germany, for example, has increased energy consumption and subsequently increased CO_2 emissions. While small decisions concerning the street or individual projects might appear minor on the regional scale, street lighting represents a major increase in energy use and CO_2 emissions. Much of the energy ends up producing skyglow and does not improve light on the ground; skyglow delineates the cities and roadways of the world from the International Space Station, as seen in Figure 5.1.

Lighting and energy designer Rob Adams in Phoenix, Arizona, outlines an apparent paradox about light at night and its energy costs. He notes the Jevons' Paradox, which the economist William Stanley Jevons outlined in 1865.[76] In this paradox, energy becomes increasingly used as it becomes cheaper, and energy expenditure actually goes up when the cost per unit goes down. Anecdotal evidence suggests that this is occurring in lighting streets. In Germany, lighting researchers are concerned at the increase in lighting because it is cheaper;[77] in Australia, new suburban areas have more poles with lights; road reconfiguration in the USA and much of the developed world now inevitably brings an increase in night-time illumination.[78] These phenomena appear to be based on old assumptions that more light is good, and that we all want it.

Not all of the world has the energy to burn lights all night. In many parts of the world, electricity is available for only parts of the day, and less of the night, or none at all. It will be important for people in less developed regions to be aware of new concerns about artificial light at night, so that they do not fall into the same assumptions that dominate the developed world. With new information they might leapfrog the developed world and avoid misguided behaviours. In contrast, the developed world might be 'stuck' with excessive lighting for decades, due to the enormous cost of making the wrong decisions today based on the single issue of energy saving, incorrect assumptions, and lack of regard for medical knowledge readily available now.

The assumptions and ubiquity of LED lighting in particular

All over the world, LED lighting is being installed with extraordinary rapidity. It has been propagated as a panacea for energy saving, and in 2014 the discoverers of LED received the Nobel Prize for Physics.[79] The Nobel Committee stated that while 'incandescent light bulbs lit the 20th Century, the 21st Century will be lit by LED lamps'.[80] By 2030 in the USA, LED will comprise 75 per cent of the market. Most decisions to convert existing systems to LED remain based on energy efficacy to the exclusion of all other factors. Early LED luminaires with a high K were more efficient than older street lights in their energy use. Thus when LED lights first began to be installed for street lighting, a typical street light was 6,000K, well in the blue range, harsh and glary; however, they were able to compete for their energy efficiency with high-pressure sodium lights. As LED efficacy improved 4,000K became the new standard. However, LEDs are not ecological or energy cure-alls, as they are so often hailed. Energy saving and being ecologically sound are not the same thing. High skyglow, increased levels of glare (a disability for driving and thus a road hazard), and an often high colour temperature (often blue-rich), all lead to concerns about ecological and human health. LED lighting also increases the impact of light pollution, irrespective of the colour temperature of the luminaire.[81] Despite these issues, LED is dominating the rapid retrofitting of the entire developed world. LED lights might well improve in the future, but there are currently many invalid assumptions and risky implications about their use[82] – assumptions of energy efficiency, inert impacts, and safety for human use.

Due to these assumptions, decision-makers rarely consult citizens during retrofitting of public lighting with LEDs In my own neighbourhood, if the local council thinks about potential changes to parking, bike lanes, traffic closure, laneway changes, or any host of ideas concerning the street and the neighbourhood, they consult their public. In the case of a complete refit of street lighting, we were told not asked, with no details of colour temperature (4,000K). Few people know what or why, and far fewer know of the implications of the lights chosen. As California's Smart Outdoor Lighting Alliance points out, the 'idea that the city chooses what the residents will have to live with is unheard of in the world of professional lighting design',[83] – where the client is consulted. The public is the client in street lighting. The client, you, has to live for decades with decisions made about public lighting without consultation, and with no actual design, despite the long life of a new street light.

What happens when communication fails between the city or local council and an increasingly better-informed public? Two examples reveal what can happen. The City of Davis in northern California planned to upgrade its lights in 2014 to save an estimated $150,000 in energy bills and maintenance. The city went ahead with the decision to upgrade with 27-watt, 2,800 luminaires (the globes) with a colour temperature of 4,000K. This is in the blue range, as was seen in Figure 5.4. Installation began in January 2014, but within a few months there were complaints that the lights were too bright, too glary, increased light trespass[84] into bedrooms, increased skyglow and light pollution, and had a potential negative impact on wildlife and human health. Although 1,400 luminaires had already been installed, the project halted. It was found that the colour temperature was the main offender, and the city decided to replace 650 lights of 4,000K with 2,700K lights (Leotek 19-watt Green Cobra™ Junior),[85] with an increase in house-side shields to reduce glare and light trespass into homes. These modifications to the original contract cost $325,000.[86] The new fixtures produce 30 per cent less illumination than the 4,000K and reduce glare for the public.

In a second case in 2013, New York City started converting 250,000 street lights to 4,000K LED but has halted.[87] Again, residents complained of the glare and cold colour temperature. The driver for this intrusion from the council was energy efficacy, ignoring research on ecology and health.[88] Currently in LED technology, 4,000K are 10 per cent more energy efficient than 3,000K or 2,700K luminaires. However, councils could just as readily think a little differently; they could install 2,700K luminaires and add dimmers to reduce the light levels significantly when most streets do not need them much – between midnight and dawn. Such adaptive controls can save far more energy than the difference between 2,700K and 4,000K. If dimmers are in operation between 10 p.m. and dawn, they could save up to 50 per cent more energy. Additionally, dimming or switching off lights after midnight would have profoundly positive impacts on wildlife and human health.

While new types of LED light are being developed that have lower colour temperature and good energy efficacy, these new lights will be too late for the millions of streetlights that have already been placed and are expected to remain in place for nearly 25 years – an entire human generation. One new 'physiologically-friendly' light has a colour temperature of 1,900K, a colour-rendering index of 93, and a high energy efficacy.[89] It is claimed to be the closest artificial light to a warm candle ever invented and is an *organic* light emitting diode (OLED).[90] There is an enormous amount of interest in this field of research, and this is set to grow as the ecological and health issues associated with LED and high K become more widely known. The American Medical Association is now encouraging an LED of less than 3,000K, strong shielding of lights, and dimming in off-peak periods.[91]

The current situation of citizens who are largely uninformed about health risks with artificial light at night will inevitably change, and when they understand, they will ask councils and legislators many questions. The main one will be: 'If the impacts of colour spectrum on human health were known' (as they now are), 'why did you not conduct best practice, or follow the precautionary principle of care?' Will local governments then be liable for human health impacted by artificial light at night?[92] Is this unacknowledged risk a sleeping giant?

Rethinking urban design for lighting at night

More light changes the nature of every place, particularly in semi-rural or urban fringe areas adjacent to nature reserves and remnant vegetation. Yet many regional hamlets and small villages are lighting in the same way as urban areas, leading to the potential loss of a sense of the rural and problems for wildlife. New research at the scales of the street and the precinct should consider the type of street and its use. For example, a busy thoroughfare to a train station will be treated differently to a quiet lane. Currently, most retro-fits of lights are implemented as if one size fits all, except for major roads.

While shielded lights lessen astronomical light pollution (light reaching the sky) by reducing light into the atmosphere, they might still cause ecological light pollution through the light's impacts on animals.[93] While a project site might have few or no biological resources, the lighting for the project might have significant adverse consequences for species adjacent to the site (via direct glare) or some distance away (via skyglow).[94] A well-known example of a redesign to prevent ecological light impacts is the story of turtles in Florida. Loggerhead turtles are severely disturbed by lights from skyglow and the indirect glare from street lighting trespassing onto the beach. Streetlights visible from the beach distracted baby turtles, so hatchlings moved in the opposite direction to the ocean. A substantial literature informs us about protecting turtle hatchlings from being disoriented by conventional street lights. 'Embedded lighting' on coastal roadways in Florida[95] removed this problem and were an effective lighting alternative for pedestrians, cyclists, and motorists.[96] While some drivers might assume that the embedded lights are too low to drive by, they are similar to runway lights suitable for aircraft travelling at take-off speed.[97]

People are becoming aware of the impacts of new, brighter lights and are seeking to ameliorate their impacts – with kinder lighting, dimming, part-night lighting, or sensor-activated lights. Dimming lights, part-night lighting, and sensors linked with computer systems allow a finer-tuning for public lighting. Local authorities are considering new procedures for dimming lights after midnight, or switch-off, or part-lighting schemes, as trials or new operational standards. While their main incentive is money saving through energy saving, there is a growing ecological and health impetus to the changes. In the UK, for example, about one-third of all local authorities have trialled dimming and part-lighting or are committed to permanent dimming and part-lighting.[98] Sixty-two English and Welsh councils, in conjunction with the police, found that neither switching off street lights, part-night lighting, nor dimming had any harmful impacts on crime or traffic collisions.[99,100]

Streets, pedestrian paths, and bike lanes are all opportunities for redesign. Replacing lamps hanging from above with embedded lighting reduces ecological impacts in parks. The Starpath is 'a sprayable coating of light-absorbing particles that harvests ultra-violet rays from the sun during the day and dramatically lights up like a starry sky at night' – the path is lit all night, but is not bright.[101] A similar night path, the Van Gogh path (Figure 5.5), responds to movement through sensors. Embedded lights have also moved to the street for cycling in Denmark.[102]

Figure 5.5 The Van Gogh path by Daan Roosengaarde in the Netherlands lights in response to the traveller through sensors.

Image: With permission from Daan Roosengaarde.

Similar to his Van Gogh footpath, artist Daan Roosengaarde and his team (in Studio Roosengaarde) in the Netherlands have devised interactive roads that glow in the dark and recharge during the day. These Smart Highways use existing road infrastructure, but are fundamentally different approaches to how we light roads. As with the idea of sensors for streetlights and starry paths, the idea is to have the road alight only when needed. Figure 5.6 shows a Smart Highway; the first one has been completed in Oss, the Netherlands, and more are anticipated. Sensor-controlled roads will have minimal light impacts on the environment and on us, and will save energy.

Motorways are also being dimmed across the world – the UK, Germany, and Australia – by accident,[103] test, or design.[104] These changes in the street point to major design possibilities for landscape architecture. The first need is a rethink of lights on poles for pathways through parks, alleys, and low-usage areas. Second is the need to work with councils, engineers, and citizens on potentials for dimming, part-night lighting, and turn-off of lights after midnight; the switch-off can be timed to correspond with the last public transport.[105] Dimming and sensor-responses (perhaps with phone apps available to local community members) require finer-tuning, public engagement, and

Figure 5.6 These highways glow in the dark and recharge during the day and would be particularly good in sun-rich environments. The road is designed to respond to the traveller. It will not be alight if there is no traffic to trip the sensors. This project is in association with Heijmans Infrastructure.

Image: https://www.studioroosegaarde.net/project/smart-highway/photo/#smart-highway. Used with permission from Daan Roosengaarde.

public education about the benefits of reduced lighting for human health and safety. This strategy constitutes a revision of assumptions. While seemingly a challenge, recent community objections to brighter LED luminaires suggests that change is coming. Chronobiology – how light affects human behaviour – is now considered inside buildings – in factories, schools, elderly care homes, and neonatal and intensive care units.[106] The next step is outside.

What we are presently missing are methods to deal with the conceptual gap between old assumptions, still tightly held, and new knowledge. The challenge for designers will be to apply this new knowledge in urban design.[107] Design disciplines and the community will share this journey into new territory. In bits and pieces, some changes are occurring, all over the world.

Our loss of the night sky is clearly noticed as part of the extinction of experience[108] of modern living. In 1988, astronomers formed the International Dark-Sky Association (IDA) with the explicit aim of protecting the night skies for future generations to experience. IDA's initiatives include International Dark Sky Parks and Dark Sky Places,[109] conservation programmes that recognise excellent stewardship of the night sky. IDA rewards stringent outdoor lighting standards and innovative community outreach with the ambition to 'encourage communities around the world to preserve and protect dark sites through responsible lighting policies and public education'.[110] Yet Dark Sky Parks commemorate dark skies as if in a museum, just like viewing the last relatives of the aurochs. Dark Sky Parks are not the solution to the questions raised in this chapter. Far harder is to address the more difficult – the lights of cities, towns, and suburbs, and to address the health and ecological impacts of current lighting regimes, lighting standards, cultural norms, and old assumptions. The more difficult project is rethinking how we light our cities. While this entire project is very much within the domain of the design and planning professions, we are largely napping amid old assumptions.[111]

Conclusions

Outlined here are compelling reasons for changes to public lighting practices. Medical scientists are calling for active work in urban design to address the problems of artificial light at night.[112] Particularly problematic are (and will be) entrenched values and institutional barriers to change, protected by top-down hierarchical organisations, including government, academia, and research institutions, that sustain 'business as usual'. Light pollution and the ubiquitous intrusion of artificial light remain low on funding priorities for sustainability or major environmental issues, largely due to assumptions maintained even in the face of new evidence that runs counter to those norms. But, as more inquiries lead to greater knowledge, practices established in the last few centuries might become maladaptive very rapidly. Designers will be key to moving to the acceptance of better choices.[113] Lights that are environmentally and medically 'better' point us towards the idea of how landscape architecture might perform for considerable social benefit[114] in an emerging arena.

Our night sky has reminded us for all of human history that our human project shelters in the universe on a small blue, green, and brown planet. The loss of our view of the night sky deprives us all living today, and also future generations. If we don't want to lose this view, we need to act now to redesign the street, the lighting methods we use, and the luminaires we choose. There is a chance to throw out the 'old street' and experiment with new forms and new lighting codes, and to avoid the extraordinary push for 4,000K lights for outdoor lighting, with blue light, all over the world.

Designers face an immense challenge to put new information about human health and ecology in urban design and planning. Currently, little design is being done with new inquiries understood. Urban designers and landscape architects can participate in this important conversation to redesign the landscapes of the night. To do so, we will need to work closely with scientists and policy-makers who understand the physics of light, the physiology of human vision, the ecological impacts of artificial light at night, medical impacts of light and spectrum, and regulations that require shifts of assumptions in response to new inquiries.

In this chapter I have cast aside concerns about light fittings, downward lights, ambience – the easy, familiar world of designed lighting and architecture and interior design. Like design georgics in the previous chapter, new information takes us toward more difficult and complex projects and toward unknown futures as a profession.

Notes

1 Pablo Neruda, 1969, 'Every Day You Play', XIV from *Twenty Love Poems and a Song of Despair*, translated into English by W.S. Merwin, New York: Penguin Books [first published 1924].
2 An auroch (*Bos primigenius*) with what appears to be the seven stars of the Pleiades star cluster above the bull's spine in the cave system of Lascaux dated at approximately 17,300 BP. The Pleiades are in what is now known as the Taurus or Bull Constellation. The cluster of dots on the cow's head correspond to the Hyades star system, the cow's eye is Aldebaran, which follows the Pleiades, and the dots to the far left correspond to the position of another star system. Although speculative, the positions of the stars suggest very strong associations.

3   Cis van Vuure, 2005, *Retracing the Aurochs: History, Morphology, and Ecology of an Extinct Wild Ox*, Sofia-Moscow: Pensoft Publishers.
4   Amelia Sparavigna, 2008, 'The Pleiades: The celestial herd of ancient timekeepers', available online at https://arxiv.org/ftp/arxiv/papers/0810/0810.1592.pdf/. Accessed 1 October 2016. She discusses the importance of the Pleiades and the associated star systems in this painting, as does Michael A. Rappenglück, 2001, 'Palaeolithic timekeepers looking at the golden gate of the ecliptic: The lunar cycle and the Pleiades in the Cave of la-tête-du-Lion (Ardèche, France) – 21,000 BP', *Earth, Moon and Planets* 85–86: 391–404.
5   Job 9:9; 38:31; Amos 5:8.
6   In Greek mythology the six brighter sisters are Maia, Elektra, Taygete, Alkyone, Kelaino, and Sterope. Merope is the name of the fainter star to the human eye. See also Robert Graves, 1955, *The Greek Myths*. Graves outlines the story that Merope had run away with her lover, who was a mere mortal, and did not want to show her face because she was ashamed of her liaison.
7   Meredith Osmond, 2007, 'Navigation and the heavens', in Malcolm Ross, Andrew Pawley and Meredith Osmond (eds), *The Lexicon of Proto Oceanic*, Vol. 2, Pacific Linguistics and ANU ePress, pp. 155–191. Along the northern coast of New Guinea, for example, the Pleiades are considered to be young unmarried women and are associated with health and fertility. And also: Jenny March, 2008, *The Penguin Book of Classical Myths*, London: Allen Lane, pp. 43–45.
8   John Edward Huth, 2013, *The Lost Art of Finding Our Way*, Cambridge, MA: Belknap Press of Harvard University Press. Huth notes (p. 134) that Polaris was not always the lodestar. While it directed Columbus, there are few if any references to Polaris in ancient times because changes in the Earth's axis lead to precession of the equinoxes, and Polaris is a 'wandering pole'. Polaris only appeared in navigation in the fifteenth century when it orbited the North Pole at an arc of four degrees. Now it has moved again.
9   The stars include the Pleiades, Altair, Orion's belt (three bright stars in a row), Sirius, Beta Scorpionis, Antares, Alpha Centauri, Canopus, and Achernar.
10 The green is oxygen and the aurora is caused by the solar wind from the Sun stretching out across the Solar System. The aurora is caused by the deflection of this high radiation dusty wind by Earth's magnetic field, which acts as a shield to the solar wind. If we lost this magnetic protection, stronger at the poles, where it is deflected, we would be exposed to harmful ultrviolet radiation.
11 Indeed, light at night is very commonly the single largest source of greenhouse gas emissions for local governments and a major target for carbon mitigation.
12 Thomas C. Erren and Russel J. Reiter, 2009, 'Light hygiene: Time to make preventive use of insights – old and new – into the nexus of the drug light, melatonin, clocks, chronodisruption and public health', *Medical Hypotheses* 73: 537–541. The authors note the need to focus on 'light hygiene'.
13 Kevin Gaston, 2013, 'A green light for efficiency', *Nature* 497: 560–561.
14 See Jane Brox, 2010, *Brilliant: The Evolution of Artificial Light*, Boston, MA: Houghton Mifflin Harcourt.
15 From an article in the *Daily Telegraph*, reporting on the Siding Springs Observatory, Canberra, Australia. 'Star wars: Lights set to be dimmed in NSW country towns to allow for space research', 27 December 2015.
16 Data from various astronomy sites and the International Dark-Sky Association.
17 This is a tyranny of small decisions, as I discuss in Grose, 2010, 'Small decisions in suburban open spaces: Ecological perspectives from a Hotspot of global biodiversity concerning knowledge flows between disciplinary territories', *Landscape Research* 35(1): 47–62.

18 Fabio Falchi, Pierantonio Cinzano, Christopher D. Elvidge, David M. Keith, and Abraham Haim, 2011, 'Limiting the impact of light pollution on human health, environment and stellar visibility', *Journal of Environmental Management* 92: 2714–2722.
19 Following a review paper: Travis Longcore and Catherine Rich, 2004, 'Ecological light pollution', *Frontiers in Ecology and the Environment* 2(4): 191–198, came the book: Catherine Rich and Travis Longcore (eds), 2005, *Ecological Consequences of Artificial Night Lighting*, Washington, D.C.: Island Press.
20 See Martin Morgan-Taylor, 2015, 'Regulating light pollution in Europe', in Josiane Meier, Ute Hasenöhrl, Katharina Krause, and Merle Pottharst (eds), *Urban Lighting, Light Pollution and Society*, New York and London: Routledge, pp. 159–176. The author gives the European legal definitions of various types of light pollution. Skyglow is the most 'pervasive form of light pollution and can affect many miles from the original light source', as noted in Royal Commission on Environmental Pollution, 2009, *Artificial Light in the Environment*, London: HM Stationery Office, p. 1.
21 Louis J. Kraus (Chair), 2016. *Human and Environmental Effects of Light Emitting Diode (LED) Community Lighting*. Report of the Council on Science and Public Health, American Medical Association. CSAPH Report 2-A-16. Accessed 1 July 2016.
22 William Brewster, 1886, 'Bird migration. Part 1. Observations on nocturnal bird flights at the light-house at Point Lepreaux, Bay of Fundy, New Brunswick', *Memoirs of the Nuttall Ornithological Club* 1, 5–10. This is an important memoir of 22 pages in all, made when Brewster observed birds on their autumn migration south.
23 Lesley J. Evans Ogden, 1996, *Collision Course: The Hazards of Lighted Structures and Windows to Migrating Birds*, Toronto: World Wildlife Fund Canada; Sidney A. Gauthreauz and Carroll G. Belser, 2006, 'Effects of artificial night lighting on migrating birds', in Catherine Rich and Travis Longcore (eds), 2005, *Ecological Consequences of Artificial Night Lighting*, Washington, D.C.: Island Press, pp. 67–93.
24 Rich and Longcore, 2005.
25 Edwin B. Matzke, 1936, 'The effect of street lights in delaying leaf-fall in certain trees', *American Journal of Botany* 23(6): 446–452. Note that Matzke found that leaves of these same trees did not emerge from buds earlier than non-illuminated trees.
26 Possibly 15 lux, presuming a tungsten incandescent light bulb.
27 They were receiving 10 lux of light at the branch – originally reported as a fraction over one foot-candle. Ibid., 447.
28 Thomas O. Perry, 1971, 'Dormancy of trees in winter', *Science* 171: 29–36.
29 Richard H. ffrench-Constant, Robin Somers-Yeates, Jonathan Bennie, Theodoros Economou, David Hodgson, *et al.*, 2016, 'Light pollution is associated with earlier tree budburst across the United Kingdom', *Proceedings of the Royal Society B* 283: 20160813.
30 See for example W. Wang, J.Y. Barnaby, Y. Tada, H. Li, M. Tör, D. Caldelari, *et al.*, 2011, 'Timing of plant immune responses by a central circadian regulator', *Nature* 470: 110–114.
31 S.L. Harmer, 2009, 'The circadian system in higher plants', *Annual Review of Plant Biology* 60: 357–377.
32 See John D. Liu, Danielle Goodspeed, Zhengji Sheng, Baohua Li, Yiran Yang, Daniel J. Kliebenstein, and Janet Braam, 2015, 'Keeping the rhythm: Light/dark cycles during postharvest storage preserve the tissue integrity and nutritional content of leafy plants', *BMC Plant Biology* 15: 92.

33 Danielle Goodspeed, John D. Liu, E Wassim Chehab, Xhengji Sheng, Marta Francisco, Daniel J. Kleibenstein, and Janet Braam, 2013, 'Postharvest circadian entrainment enhances crop pest resistance and phytochemical cycling', *Current Biology* 23: 1235–1241.
34 A. Outen, 1998, *The Possible Ecological Implications of Artificial Lighting*, Hertford: Hertfordshire Biological Records Centre.
35 S. Sinnadurai, 1981, 'High pressure sodium street lights affect crops in Ghana', *World Crops* 33(6): 120–122; H.M. Cathey and L.E. Campbell, 1975, 'Effectiveness of five vision-lighting sources on photoregulation of 22 species of ornamental plants', *Journal of the American Society for Horticultural Science* 100: 65–71.
36 Gerhard Eisenbeis (2005) reports work in Germany, where 3 million insects were killed every night, in Catherine Rich and Travis Longcore (eds), 2005, *Ecological Consequences of Artificial Night Lighting*, Washington, D.C.: Island Press, pp. 281–304.
37 See for example, recent work that shows a reduction in firefly populations in Brazil with artificial night lighting: Oskar Hagen, Raphael M. Santos, Marcelo N. Schlindwein, and Vadim R. Viviani, 2015, 'Artificial night lighting reduces firefly (Coleoptera: Lampyridae) occurrence in Sorocaba, Brazil', *Advances in Entomology* 3(1): 24–32.
38 With the Tribute in Light, Ground Zero, New York, many birds moving along the Atlantic Flyway have become trapped in this blue light, and the lights were turned off whenever the numbers of birds exceed 1,000. That is a great many birds. Their migration coincides with the memorial day for 9/11. Details from Audubon Society of America and http://www.dailymail.co.uk/news/article-3232091/Tribute-Light-shut-four-times-THOUSANDS-migrating-birds-trapped-like-moths-powerful-beams-produced-Earth-commemorate-victims-9-11.html.
39 Kevin J. Gaston, Thomas W. Davies, Jonathan Bennie, and John Hopkins, 2012, 'Reducing the ecological consequences of night-time light pollution: Options and developments', *Journal of Applied Ecology* 49: 1256–1266.
40 For tammar wallabies (*Macropus eugenii*) exposed to artificial light at night, birthing peaks in February, a month later than normal in those animals that experience dark bushland at night. Mothers with joeys born too late face food shortages. The reproductive cycle in tammar wallabies is synchronised with day length. There is growing evidence that the female of any species is more greatly impacted by light at night than the male.
41 Kylie A. Robert, John A. Lesku, Jesko Pertecke, and Brian Chambers, 2015, 'Artificial light at night desynchronizes strictly seasonal reproduction in a wild mammal', *Proceedings of the Royal Society B* 282: 20151745.
42 Joshua J. Gooley, Kyle Chamberlain, Kurt A. Smith, Sat Bir S. Khalsa, Shantha M.W. Rajaratnam, *et al.*, 2011, 'Exposure to room light before bedtime suppresses melatonin onset and shortens melatonin duration in humans', *Journal of Clinical Endocrinology and Metabolism* 96(3): E463–E472. The authors found that 'room light exerts a profound suppressive effect on melatonin levels and shortens the body's internal representation of night duration'. This occurred in 99 per cent of all people tested. Room light shortened the body's internal representation of night duration. The authors note that the disruption of melatonin signalling the night could impact sleep, thermoregulation, blood pressure, and glucose homeostasis. Fundamentally, exposure to light resets the circadian rhythm of melatonin and inhibits its synthesis.
43 Gena Glickman, Robert Levin and George C. Brainard, 2002, 'Ocular input for human melatonin regulation: Relevance to breast cancer', *Neuroendocrinology Letters* 23 (suppl. 2): 17–22.

44 Personal comm.
45 Dennis M. Dacey, His-Wen Liao, Beth B. Peterson, Farrel R. Robinson, Vivianne C. Smith, Joel Pokorny, King-Wai Yau and Paul D. Gamlin, 2005, 'Melanopsin-expressing ganglion cells in primate retina signal color and irradiance and project to the LGN', *Nature* 433: 749–775.
46 S. Hattar, H-W Liao, M. Takao, D.M. Berson, and K-W Yau, 2002, 'Melanopsin-containing retinal ganglion cells: Architecture, projections, and intrinsic photosensitivity', *Science* 295: 1065–1070.
47 Dingcai Cao and Pablo A. Barrionuevo, 2015, 'The importance of intrinsically photosensitive retinal ganglion cells and implications for lighting design', *Journal of Solid State Lighting*, available online at doi:10.1186/s40539-015-0030-0/ Accessed 12 February 2016.
48 Two important papers on this are: C.S. Pechacek, M. Andersen, and S.W. Lockley, 2008, 'Preliminary method for prospective analysis of the circadian efficacy of (day) light with applications to healthcare architecture', *Leukos* 5: 1–26. And: M. Andersen, J. Mardaljevic, and S.W. Lockley, 2012, 'A framework for predicting the non-visual effects of daylight – Part 1: Photobiology-based model', *Lighting Research and Technology* 44: 37–53.
49 As noted by J.S. Cha, J.W. Lee, W.S. Lee, J.W. Jung, K.M. Lee, J.S. Han and J.H. Gu, 2014, 'Policy and status of light pollution management in Korea', *Lighting Research and Technology* 46: 78–88.
50 The international unit of length for one-billionth of a metre is nm = nanometre.
51 Blue light in LED screens can suppress melatonin and increase alertness, and can impair sleep. Orange-tinted blue light-blocking glasses (named blue-blockers) can be used to filter out the blue range of the spectrum; trials of this prevented the suppression of melatonin in young adults. Blue light is a growing field of investigation: see, as examples, Christian Cajochen, Sylvia Frey, Doreen Anders, Jakub Späti, Matthias Bues, Achim Pross, Ralph Mager, Anna Wirz-Justice, and Oliver Stefani, 2011, 'Evening exposure to a light-emitting diodes (LED)-backlit computer screen affects circadian physiology and cognitive performance', *Journal of Applied Physiology* 110(5): 1432–1438; they note that 'the challenge will be to design a computer screen with a spectral profile that can be individually programmed to add timed, essential light information to the circadian system in humans'. See also: Brittany Wood, Mark S. Rea, Barbara Plitnick, and Mariana G. Figueiro, 2013, 'Light level and duration of exposure determine the impact of self-luminous tablets on melatonin suppression', *Applied Ergonomics* 44: 237–240. For background to the blue-blockers, see: Stéphanie van der Lely, Silvia Frey, Corrado Garbazza, Anna Wirz-Justice, Oskar G. Jenni, Roland Steiner, Stefan Wolf, Christian Cajochen, Vivien Bromundt, and Christina Schmidt, 2015, 'Blue blocker glasses as a countermeasure for alerting effects of evening light-emitting diode screen exposure in male teenagers', *Journal of Adolescent Health* 56(1): 113–119.
52 According to Australian local government councils.
53 William C. Dement and Christopher Vaughan, 1999, *The Promise of Sleep: A Pioneer in Sleep Medicine Explores the Vital Connection between Health, Happiness, and a Good Night's Sleep*, New York: Dell Publishing Co., pp. xiii.
54 Early work is in: Drew Dawson and Kathryn Reid, 1997, 'Fatigue, alcohol and performance impairment', *Nature* 388: 255; the authors suggest an equivalent 'blood alcohol reading' for sleep deprivation; Jane C. Stutts, Jean W. Wilkins, and Bradley V. Vaughn, 1999, 'Why do people have drowsy driving crashes? Input from people who just did', AAA Foundation for Traffic Safety online, accessed 28 April 2016; A.M. Williamson and Anne-Marie Feyer, 2000, 'Moderate sleep deprivation produces impairments in

cognitive and motor performance equivalent to legally prescribed levels of alcohol intoxication', *Occupational and Environmental Medicine* 57: 649–655.

55 Christian Vollmer, Ulrich Michel, and Christoph Randler, 2012, 'Outdoor light at night (LAN) is correlated with eveningness in adolescents', *Chronobiology International* 29(4): 502–508. Shantha Rajaratnam at Monash University has noted to me that we once thought that people had sleep disorders due to psychological problems, but it is realised now that many people have mental health problems due to sleep disturbance.

56 Martin Forejt, Jan Hollan, K. Skočovsky, and R. Skotnica, 2004, 'Sleep disturbances by light at night: Two queries made in 2003 in Czechia', poster at Cancer and Rhythm, Graz, Austria. Online. This preliminary study in central Europe suggests an interesting wider study of waking preference.

57 Tara A. LeGates, Diego C. Fernandez and Samer Hattar, 2014, 'Light as a central modulator of circadian rhythms, sleep and affect', *Nature Reviews. Neuroscience* 15: 443–454.

58 T.S. Wiley and Bent Formby, 2000, *Lights Out: Sleep, Sugar, and Survival*, New York: Pocket Books, sets out discussion in an easily read form. More information has since come to light in medical science. For a recent review see: J. Cipolla-Neto, F.G. Amaral, S.C. Afeche, D.X. Tan, and R.J. Reiter, 2014, 'Melatonin, energy metabolism, and obesity: A review', *Journal of Pineal Research* 56: 371–381; the authors discuss chrono-disruption leading to obesity.

59 In association with this has been the finding that mean sleep duration in the US population has decreased by one to two hours over the past 40 years.

60 Richard G. Stevens, 2009, 'Electric light causes cancer? Surely you're joking, Mr Stevens', *Mutation Research* 682: 1–6.

61 My own experience of this was when a collaborator, Clare Mouat, a planner, presented the link at a planning conference; she was asked if our research was being funded by muggers and rapists. Such comments do nothing to advance discussion, but show how entrenched assumptions can be.

62 In lecturing on this subject over the last few years in Melbourne, I recommend to students that, if they wake during the night, they do not look at mobiles, turn on a light, or open the fridge – or their melatonin will crash.

63 See for examples: Scott Davis, Dana K. Mirick and Richard G. Stevens, 2001, 'Night shift work, light at night, and risk of breast cancer', *Journal of the National Cancer Institute* 93(2): 1557–1562. A recent review is the book by Abraham Haim and Boris A. Portnov, 2013, *Light Pollution as a New Risk Factor for Human Breast and Prostate Cancers*, Dordrecht and New York: Springer.

64 Itai Kloog, Abraham Haim, Richard G. Stevens, Micha Barchana, and Boris A. Portnov, 2008, 'Light at night co-distributes with incident breast but not lung cancer in the female population of Israel', *Chronobiology International: The Journal of Biological and Medical Rhythm Research* 25(1): 65–81.

65 Richard G. Stevens, George C. Brainard, David E. Blask, Steven W. Lockley and Mario E. Motta, 2014, 'Breast cancer and circadian disruption from electric lighting in the modern world', *CA: A Cancer Journal for Clinicians* 64: 207–218.

66 Council on Science and Public Health Report 4. *Light Pollution: Adverse Health Effects of Nighttime Lighting*. American Medical Association House of Delegates Annual Meeting, June 2012, Chicago, USA.

67 Stephen M. Pauley, 2004, 'Lighting for the human circadian clock: Recent research indicates that lighting has become a public health issue', *Medical Hypotheses* 63: 588–596.

68 Margaret J. Grose, 2014, 'Artificial light at night: A neglected population health concern of the built environment', *Health Promotion Journal of Australia* 25: 193–195.

69 Richard G. Stevens, 2009, 'Light-at-night, circadian disruption and breast cancer: Assessment of existing evidence', *International Journal of Epidemiology* 38: 963–970.
70 BBC News, 29 July 2015, *Less lighting has no impact on crime or collisions, says report*, http://www.bbc.com/news/uk-33692675. This BBC report based on Rebecca Steinbach, Chloe Perkins, Lisa Tompson, Shane Johnson, Ben Armstrong, *et al.*, 2015, 'The effect of reduced street lighting on road casualties and crime in England and Wales: Controlled interrupted time series analysis', *Journal of Epidemiology and Community Health* 69(11): 1118–1124.
71 Ibid.
72 Bob Mizon, 2012, *Light Pollution: Responses and Remedies*, New York: Springer, pp. 85–97. See also: Stephen Atkins, Sohail Husain and Angele Storey, 1991, 'The influence of street lighting on crime and fear of crime', Crime Prevention Unit Paper no. 28, London: Home Office; they found no evidence that increased light decreases crime.
73 Mizon, 2012.
74 P.R. Boyce, N.H. Eklund, B.J. Hamilton and L.D. Bruno, 2000, 'Perceptions of safety at night in different lighting conditions', *Lighting Research and Technology* 32(2): 79–91.
75 Royal Commission on Environmental Pollution, 2009, *Artificial Light in the Environment*, London: HM Stationery Office, p. 1.
76 William Stanley Jevons, 1865, *The Coal Question: An Inquiry Concerning the Progress of the Nation, and the Probable Exhaustion of Our Coal Mines*, London: Macmillan and Co. Jevons argued that greater economy increased consumption, leading to greater use of the resource. He also noted the responsibility of a current generation to future generations, foreshadowing the precautionary principle of sustainability.
77 Christopher C.M. Kyba, Andreas Hänel, and Franz Hölker, 2014, 'Redefining efficiency for outdoor lighting', *Energy & Environmental Science* 7: 1806–1809.
78 Travis Longcore and Catherine Rich, 2001, *A Review of the Ecological Effects of Road Reconfiguration and Expansion on Coastal Wetland Ecosystems*, Urban Wildlands Group, California, at http://www.urbanwildlands.org/conference.html/ Accessed 19 February 2016.
79 The Nobel Committee cited Isamu Akasaki and Hiroshi Amano and Shuji Nakamura for 'the invention of efficient blue light-emitting diodes, which has enabled bright and energy-saving white light sources'.
80 International Dark-Sky Association 2014, *IDA responds to 2014 Nobel Prize for Physics*, available online at http://darksky.org/2014-nobel-prize-for-physics-draws-attention-to-promise-and-challenges-of-blue-light/ Accessed 29 December 2015.
81 Steve M. Pawson and M.K.F. Bader, 2014, 'LED lighting increases the ecological impact of light pollution irrespective of colour temperature', *Ecological Applications* 24: 1561–1568.
82 Robert Adams, C and W Energy Solutions Phoenix, Arizona. On IDA website, November 2015. Adams outlines the myths that LED lights reduce light pollution because they are energy efficient, that LED lights are easier to control in terms of where light lands on the ground; that they increase traffic safety; that they improve security by discouraging crime; and that LED means a lower carbon footprint.
83 David Vanguard, 2015, at http://volt.org/why-color-matters-designing-for-the-real-client/ Accessed 20 December 2015.
84 Light trespass is unwanted light, such as from adjacent properties or streetlights.

85 The manufacturers note that the Leotek 2,700 LED is the most 'unchallenged' LED street light in the world – that is, it has been the most acceptable to the public.
86 Available online at http://volt.org/lessons-learned-davis-ca-led-streetlight-retrofit/; and http://sacramento.cbslocal.com/2014/10/21/davis-will-spend-350000-to-replace-led-lights-after-neighbor-complaints/ Accessed 22 December 2015.
87 Reported by Miranda Katz, 2016, *City Will Replace Blinding LED Streetlights*, available online at http://gothamist.com/2016/05/14/the_city_is_replacing_those_blindin.php/ Accessed 13 July 2016.
88 In my own city the push to replace interior lights with LED has been relentless but with no discussion of colour temperature and spectrum, no noting of anything whatsoever but money. The advertising for such changes feature the words 'eco', care, carbon saving, energy efficiency, money saving, and free, and promise cheaper electricity bills. The assumption is that LED will be our saviour.
89 Jwo-Huei Jou, Chun-Yu Hsieh, Jing-Ru Tseng, Shiang-Hau Peng, Yung-Chen Jou, *et al.*, 2013, 'Candle light-style organic light-emitting diode', *Advanced Functional Materials* 23(21): 2750–2757.
90 Organic LEDs are based on the use of an organic substance – small organic molecules in a crystalline phase, or often a polymer – as the semi-conductor material.
91 Kraus, 2016.
92 If the community feels powerless to complain about new lights, or if the council is too cash-strapped to be able to change and the community would know that, no backlash might occur immediately. Another issue here is environmental generational amnesia. If our lights remain bright at night, and all night, our grandchildren may grow up inexperienced in walking in dark conditions, or on darkish streets, and become afraid of the night, despite the many benefits of darkness.
93 The distinction between astronomical light pollution and ecological light pollution is that the astronomical is purely concerned with light being sent upward to act as skyglow. Ecological light pollution can occur with shielded lights or dim lights (which might prevent astronomical light pollution) if the lights are disruptive of animal behaviour.
94 Catherine Rich and Travis Longcore (eds), 2006, *Ecological Consequences of Artificial Night Lighting*, Washington, D.C.: Island Press.
95 Michael Salmon, 2006, 'Protecting sea turtles from artificial night lighting at Florida's oceanic beaches', in Catherine Rich and Travis Longcore (eds), *Ecological Consequences of Artificial Night Lighting*, Washington, D.C.: Island Press, pp. 141–168.
96 Ibid.
97 In Australia, alternative street lighting is used in Port Hedland at suburban Pretty Pool to protect a turtle hatchery on an adjoining beach. Design by ERM, a landscape architectural firm in Perth, Western Australia. Information from Greg Grabash, pers. comm.
98 The Campaign to Protect Rural England collected this information. *Shedding Light: A Survey of Local Authority Approaches to Lighting in England*. April 2014. Available online at www.cpre.org.uk.
99 Ibid., pp. 20–21. The authors note that 91 per cent of councils that are switching off are monitoring crime, while only 51 per cent of councils that are dimming streetlights are doing so.
100 Rebecca Steinbach, Chloe Perkins, Lisa Tompson, Shane Johnson, Ben Armstrong, Judith Green, Chris Grundy, Paul Wilkinson, and Phil Edwards,

2015, 'The effect of reduced street lighting on road casualties and crime in England and Wales: Controlled interrupted time series analysis', *Journal of Epidemiology and Community Health* 69: 1118–1124.

101 Neil Blackmore: ProTeq marketing agent; www.Pro-Teqsurfacing.com; info@Pro-Teqsurfacing.com. As local councils are increasingly cutting off park lights at night to save money, StarPath guides the walker or cyclist. The non-reflective surface does not give off skyglow. Available online at http://www.takepart.com/article/2013/10/30/starpath-glow-in-the-dark-roads-provide-energy-free-illumination; see also http:// www.forumforthefuture.org/greenfutures/articles/glow-dark-%E2%80%98starpath%E2%80%99-lights-cambridge-park/ Accessed 19 December 2015.

102 Embedded lighting in Copenhagen has put green lights in a bicycle path that runs on a major artery into the city. 'Lighting the way to a green city: Copenhagen's Smart Streets reduce Energy Use'. Reported in the *New York Times*, 9 December 2014. The title of the article suggests that energy was the sole focus. However, it is a simple change and effective to make this type of alteration in the design of the street.

103 In Perth, Western Australia, engineers had to dim and part-light a section of a major freeway due to an energy disruption. The reduced lighting received no negative feedback from drivers, and the situation remains, long after the energy disruption ended.

104 Dimming motorways has also been found to reduce ocular stress.

105 Margaret J. Grose, Clare Mouat, and Doug Ayre, 2012, 'Enlightened urban transformations: Reflecting on the beginnings of street lighting in Perth', Proceedings of the 11th Australasian Urban/Planning History Conference: *Urban Transformations: Booms, Busts and Other Catastrophes*, Perth, University of Western Australia. Many cities used to turn off streetlights when the last tram, bus, or train had finished, with a few minutes added either side.

106 Mirjam Münch and Vivien Bromundt, 2012, 'Light and chronobiology: Implications for health and disease', *Dialogues in Clinical Neuroscience* 14: 448–453.

107 See: International Dark-Sky Association's 2012 book: *Fighting Light Pollution: Smart Lighting Solutions for Individuals and Communities*, Mechanicsburg, PA: Stackpole Books.

108 Robert M. Pyle first used this expression in 1978: and in *The Thunder Tree*, 1993, Eugene, OR: Oregon State University Press. It has been used widely since in the fields of play, psychology, and design; less so in ecology.

109 For more information, visit: darksky.org/night-sky-conservation/dark-sky-places; for an analysis see: Josiane Meier, 2015, 'Designating Dark Sky areas', in Josiane Meier, Ute Hasenöhrl, Katharina Krause, Merle Pottharst, *Urban Lighting, Light Pollution and Society*, London: Routledge, pp. 177–196.

110 International Dark-Sky Association. Available online at http://darksky.org/idsp/ Accessed 9 February 2016.

111 There is a general need for scientific evidence on the effects of artificial light on circadian rhythmicity and its impacts on human health to be applied in public health and local government policies.

112 A. Haim and B.A. Portnov, 2013, 'Dark-less world: What is next? (Conclusions and prospects for future research)', in A. Haim and B.A. Portnov, *Light Pollution as a New Risk Factor for Human Breast and Prostate Cancers*, Dordrecht: Springer Science+Business Media, pp. 139–144.

113 Frans Berkhout, Julia Hertin, and David M. Gann, 2006, 'Learning to adapt: Organisational adaptation to climate change impacts', *Climate Change* 78: 135–156. The authors discuss the potential of change in terms of

organisational learning and potentials for accepting innovation, both key issues in lighting because lighting policies are directed by local government; their capacity for adaptation is essential for improvements in public lighting.

114 The idea of performance assessment for social benefits is discussed in: Bo Yang, Shujuan Li, and Chris Binder, 2015, 'A research frontier in landscape architecture: Landscape performance and assessment of social benefits', *Landscape Research*, available online at doi:10.1080/01426397.2015.1077944.

6

THINKING BACKWARDS, NOT FORWARDS AS A LINEAR NARRATIVE

> Aber diese gepriesene 'Philologische Methode' ... gibt es so wenig wie eine Methode, Fische zu fangen. Der Wal wird harpuniert, der Hering im Netz gefangen. Der Butt wird getreten, der Lachs gespiesst, die Forelle geangelt. Wo bleibt da *die* Methode, Fische zu fangen?
>
> (Why, this prized 'philological method'? There simply isn't any – any more than a method to catch fish. The whale is harpooned; the herring caught in a net; flounders are stomped upon; the salmon speared; the trout caught on a fly. Where do you find *the* method to catch fish?)[1]

Recently, a student in a design studio came to see me because he was 'stuck'. His site in Banda Aceh adjoined the Aceh Tsunami Museum, built in 2009 as a memorial to the Indian Ocean Tsunami of December 2004; the site included seventeenth-century Dutch gardens and a fruiterers' paradise. Having survived the tsunami himself and being a local, he knew the area well, but his wad of site analysis had made him lose sight of his original excellent ambition for the area. He himself recognised the problem as searching for a solution through a linear process of data analysis and mapping assessments, leaving him stuck in the details of the problem-description phase. Some students refer to this as 'analysis paralysis'. The issue here is about data – how we collect data and how we use it in design.

The push to collect more and more data is neither new nor unique to design. The mathematician and philosopher Jean de Rond d'Alembert (1717–1783), one of the French Encyclopaedists of the eighteenth century, described the same dangers of excessive collection of data and facts in his *Preliminary Discourse to the Encyclopédie*:

> The world of erudition and of facts is inexhaustible; the effortless acquisitions made in it lead one to think that one's substance is

continually growing, so to speak. On the contrary, the realm of reason and of discoveries is rather small.[2]

Teachers and students face the trap of thinking that more and more data for site analysis will reach an end point where an enlightened answer will emerge from the 'substance' of accumulated data. Fundamentally, the problem lies in thinking that design is a linear process, with a beginning and a series of additions of knowledge from data, leading to an end. In this chapter I address the need to approach data in a non-linear manner, and to avoid the seduction of the mere accumulation of 'substance', with neither enlightenment nor strong design outcomes.

How we deal with data is important because of its increasing role in design, and the emerging role of Big Data (and with it, the coming data deluge). A looming challenge for designers is to deal with more data than ever before experienced. Within constructing ecologies, we need to engage with increasing data more explicitly and to a greater degree because of the field's strong links with the physical and biological sciences, and with environmental flows and states. How will we as designers deal with this data? What thinking is involved, and what role do current and emerging technical and digital capacities play to enable designers to best deal with data?

Science reveals something to us of the shift away from linear thinking. Science has moved from framing many processes as linear, or forward, processes. Most well known in botany is the idea of succession leading to a final climax state. This is now no longer seen as the case, though it is still a helpful idea in restoration. In the same way as ecologists now see systems as more fluid and changeable, with fuzzy and fluxing histories, now design is moving into deliberate testing of fluid permutations of possibilities. With this shift in thinking, there will likely come a better understanding by designers about the nature of science as practised, which lies well beyond mere empiricism and exists equally in the realms of testing, ideas, queries, imagination, framing the question,[3] and probing how to test the questions for variable solutions. Many future practices will operate in a shared manner of working, and working together, with science and design, in order to find solutions to the questions and difficulties raised in constructing ecologies. What thinking is involved?

Here, I outline that there is no one method of reaching an outcome, no linear route. However, there are multiple possibilities of what can be done at a site to construct an ecologically sound outcome, and it is often best to start at the ambition and work backwards. While designers often follow this in practice, I set out this idea theoretically.

Getting out of a problem by addressing the real problem

The excellent student from Indonesia I described above is not alone in the world, as both students and teachers will attest. In their book on design teaching, *Design Expertise*, Bryan Lawson and Kees Dorst noted a design experiment during a short charrette given to both second-year novices and final-year industrial design students. The second-year students did better because, it transpired, they saw a clear problem little hampered by complexity and playfully tested designs. In contrast, the final years spent a great deal of time on the design analysis and got caught with their own analytics.[4] This is not unique. Over the last few decades, design has often been taught as growing out of site analysis and opportunities and constraints in the site.

Teaching arising from the 'overlay technique' has dominated site analysis in design studio since the American landscape architect Ian McHarg published his seminal book *Design with Nature* in 1969.[5] The overlay technique essentially recommends establishing a biological partnership with nature. While *Design with Nature* is one of the most influential books of the twentieth century in landscape architecture, for many years designers have quietly criticised McHarg's technique because of the prescriptive nature of his idea and its method, although prescription was not his intention at all. However, the idea of overlay mapping has led to a lineage of teaching that stresses an essentially linear form of analysis towards design. The overlay technique spatially located components or elements of a site and suggested a sifting of this spatial information to reveal a design solution to the site. The technique suggests that an answer is reached if all of these assembled components are considered. However, the overlay technique cannot deal with the processes between the systems it is mapping. This is an essential problem when we consider constructing ecologies because it defies the systems thinking that is the fundamental core of ecology or natural processes; we cannot test how interactions occur through the two-dimensional overlay technique. It is more useful for examining spatial layouts for planning, for which it can give some guidance, but in design the overlay technique becomes a two-dimensional reading of a three-dimensional world and is, therefore, inherently limited.

Often, to many students, the procedure of the overlay technique appears to set up a 'functional fixedness' in approaches to the design site. Functional fixedness is using a tool in a manner already described or as expected, and this fixedness can be set very early in our lives. For example, children as young as about 6 'come to possess artefact concepts that embody the function the artefact was designed to perform as a core property';[6] in other words, they are limited by how they use something because its normal manner of use has already become set in them. In landscape design, functional fixedness can be considered as the methodology used to obtain an understanding of how to design. Research in functional fixedness suggests that knowledge of an existing or fixed use or method constrains an attempt to find a solution creatively.[7] This research suggests that a procedure such as the overlay technique gives too much direction and pattern to achieve good design and is likely to lead to great difficulty to create anything atypical – killing dynamism, intuition, reinvention, the 'grand gesture', and flair.

Two-dimensional spatial techniques confine our design to Flatland and do not allow an inquiry into processes and interactions that occur in three dimensions. Flatland is an expression I am borrowing from a curious and theoretically helpful little masterpiece written by the English clergyman and educator, Dr Edwin Abbott, in 1884. *Flatland: A Romance in Many Dimensions*[8] is a social satire that describes the journeys into all the four dimensions of Mr A. Square, who is normally a resident of two-dimensional Flatland, where women are segments of lines and men are polygons or circles, depending on class. When Mr A. Square, a mathematician, visits one-dimensional Lineland, he finds life to be intolerable because everyone is a line, and if a line approaches you head on, all you see is a point, which is rather frightening. Then a Mr Sphere takes Mr Square into Spaceland, where Square discovers the joys of three dimensions. Importantly for my story, Mr Square is able to have relationships in three dimensions, while he found this impossible in two dimensions. The need to be out of Flatland is a key to constructed ecologies because shifting thinking out of two-dimensional mapping and moving to working in three dimensions enables designers to address processes and interactions between processes, as well as human and other animal behaviours. A flat plane of two dimensions cannot deal with complexity or with ecology, because every ecology lives in three-dimensional space and in the fourth dimension of time (as seen in Chapter 2). Every constructed ecology does the same, and any inquiry into ecology demands thinking beyond mapping techniques in two dimensions.

Current changes in our thinking and in our capacities for testing and experimenting are moving design away from methodologies based on two-dimensional techniques and mapping. We are beginning to re-examine sites in the light of performance and systems thinking, using three-dimensional digital capacities to interrogate designs and sites. Digital tools enable us to work with processes and precise information. The change in interrogation raises core issues of how we approach design intellectually – what are we actually doing and thinking? Such questions are important, as we are now able to test ideas while they are still in the plastic form before being constructed.

Two-dimensional thinking also runs the danger of restraining us theoretically in the problem-description phase. This danger is important for design because virtually all of what designers do is about actions and outcomes; that is, the problem is described – often by science – and we answer the question posed by the site. Much of science is concerned with data collection and assessment to ascertain hypotheses and trends that can then inform decisions, and designers rely on science to inform their design decisions. While both science and design suffer from summaries, simplification, and representational problems of data within the realm of problem-description, data from science are particularly important in constructed ecologies, as data present detail, suggest likelihood of success, and convey knowledge of processes framed within the design. Data are essential – in order to examine or test the likely performance of the designed site, or the systems characteristics of the site at any stage. That is, everything that design creates involves action.

Thinking forwards and thinking backwards

McHarg's overlay technique was essentially placing design as a forward problem, where we know all of the components and will use them to discover a solution. A forward problem in mathematics is defined as one where we do not know the answer, but have all or most of the components of the question. We deal with forward problems all the time, particularly when we are at school. Fundamentally, we all know the answer to $2 + 2 =$?; this is a forward problem. Because all of the components of a forward problem are locked into place, a forward problem typically has only one answer; in a normal, forward problem we arrive at a single correct answer, and in mathematics other answers will be incorrect. Ian McHarg never intended the overlay technique to be prescriptive, but for it to serve as a manner of approaching the understanding of a site, albeit in a forward way. Indeed, McHarg hinted at the ability to use the overlay technique as an inventory for interacting processes.[9] However, due to the manner in which they have been taught, the overlay and other mapping techniques have reinforced design as a forward problem. Due to its ease of technique and ease of teaching, the overlay technique has become a standard process of description and annotation of site analysis, and not an inquiry where we can deal with complex, interactive ecologies, whether newly constructed or under repair. This has been an error.

One of the problems with design as a forward problem is that, in teaching, it is very easy for students to be overwhelmed or seduced by site analysis: more and more data seems better when it is not – and this can delay their development as a designer and can set patterns on the ground that dictate a limited direction of thought, or no thought at all. Often, data collection remains as just more and more describing the site, not reaching decisions towards an action. A further phenomenon of site analysis in the forward-problem manner is that most students find it hard to move their design in different directions. Another problem with forward thinking for design can be considered in the light of Ken Robinson's comment on the nature of creativity and teaching for creativity: 'Teaching for creativity involves asking open-ended questions where there may be multiple solutions.'[10]

I am suggesting that we are now doing something quite different in the way we design, particularly when we design constructed ecologies. We are thinking backwards. That is, we can take design as an inverse problem.[11] Simply put, we know what our solution is, because our solution is fundamentally an ambition. With inverse problems we know the answer – in our case, it is a well-functioning ecology that we are to build, but need to work out how to construct it. By seeing design as an inverse problem and thinking backwards, we can get out of the problem-description phase; we can shift from data collection – where we collect information, patterns, models, and distributions – to one where we know the desired solution, such as clean storm-water or increased biodiversity. We can then work backwards to test what variations in the components of the system might achieve the desired outcome.

Backcasting is a term also used in ecology and sustainability.[12] In backcasting, a future is imagined; however, in backcasting one moves backwards from a desired sustainability idea, step by step, to work out how to get from that desired state to where we are today, and then a plausible causal chain of action is proposed. Like inverse thinking, backcasting moves us away from forward thinking – 'forecasting', and in doing so, it moves us away from projecting today's issues and assumptions onto the future.[13] Unlike forecasting, backcasting is not intended to predict the future. However, backcasting can establish key objectives or behavioural changes needed to get to a possible future. Stretch goals in business and management are ambitious long-term goals designed to inspire creativity and innovation. Stretch goals are similar to thinking backwards in that they create a vision to work towards that is often a tangible entity; however, in contrast, they are designed 'to drive the pursuit of answers to problems to which the solution is currently unknown'.[14] Elasticity analysis is another technique used by ecologists in population management to provide information as to which life stages of an organism have the greatest impacts on a population, or on the predators of that population. This technique helps managers to locate the 'Achilles heel' of an organism and lets them focus their management energies on that crucial area.[15] All of these ideas work with the concept of knowing, in some way, the 'answer' and looking backwards not forwards.

Approaches concerning inverse problems for design

Inverse thinking for design can be a more dynamic process than forward problems because we can test all sorts of possibilities to solve our design questions. Exactly what is an inverse problem? That an inverse problem is not a normal problem is a tautology, but it helps to consider, first, what a normal problem might be. Ideas of both normal and inverse problems come to us straight from mathematics. A normal problem in mathematics is one where the answer is computed from components of the problem; you do not have the answer when you start. A typical and easy normal problem was given as $2 \times 2 = 4$. In $3 \times 3 = ?$ and $3.2 \times 3.127 - 1.0064 = ?$, the answer is 9 in both cases. In these normal problems the answer is not stable to small changes in inputs; thus 3.0001×3 does not give 9 as an answer. That is, and put simply, the data in a normal or forward problem give a single, immutable answer.

In contrast to a normal problem, in an inverse problem the final solution is known. In an inverse problem, we work backwards to see how to reach the answer, and the ways of arriving at the answer in design might be many and varied, dependent on methods, sites, and a host of complex interactions and decisions. In contrast to the instability of the answer in normal problems – that is, the answer is changed with tiny changes in the parts as we see in 3.0001×3 – inverse problems are stable to small changes in inputs, because the desired answer is mutable. Mathematicians discuss inverse problems in terms of non-unique solutions. We can see this clearly in our ecological ambitions for a site; there are many ways to get to the 'answer'; some ways will be subtly different, and some considerably different from one another; the answers themselves vary while under the same ambition. This is clearly how designers of landscape and constructed ecologies generally think.

Examples of inverse problems abound from both ecology and design, and they are found too in constructed ecologies. Daylighting rivers is a classic inverse problem, where the strong ambition is to uncover the river; everything else follows that ambition. We can see this clearly in Cheonggyecheon in Seoul, and in the Sawmill River in Yonkers, New York, where the ambitions were to daylight misused rivers that had been covered in order for transportation infrastructure to be built above them (Figure 6.1). With inverse problems, the solutions become our ambition or our 'result' – the end point to where we are going. In this way, we know the ambition before we have the components of the 'equation' to get there.

Inverse problems are considered to be ill-posed questions and have been of interest for a century, mainly because they occur widely.

Figure 6.1 Cheonggyecheon in Seoul, completed in 2005, reveals an ambition of restoring a river and daylighting a city stream. To achieve this, designers and planners worked backwards from the ambition.

Image: From Wikipedia Commons.

Ill-posed questions violate one or more of the criteria for well-posed questions – that the problem must have a solution, the solution must be unique, and the solution must be stable under small changes to the data.[16] In the early twentieth century Henri Bergson and other process philosophers, such as Alfred North Whitehead, first discussed inverse problems and thinking backwards.[17] The American logician and mathematician Charles Sanders Peirce had already discussed these ideas, although a professional nemesis had suppressed his work throughout this life. Peirce wrote about the need 'to know what we think, to be masters of our own meaning'[18,19] in order to articulate exactly how we think, in order to think better. The idea of inverse problems has taken many slightly varying forms and has been both useful and critical across many disciplines, including tomography, civil engineering, and geophysics. Inverse problems fall into two broad categories of parameter estimation, and design optimisation.[20] In design optimisation, the desired outcome is used to solve the problem or question posed, such as a good ecological function with cultural and aesthetic qualities. Engineer and designer Kees Dorst has referred to the 'solution' in an inverse problem as an 'aspired value' or an aspiration, which is a fine name for the ambition of our designs[21] – that might be a subsurface wetland or a new public park.

Mathematics can describe inverse problems simply.[22] If we think of a digital camera, it makes an image by attempting to recreate the true scene (**X**), but compromises the perfect image, creating instead the image **Y** that is distorted due to the presence of noise ε (epsilon). Thus, for scene **X** and its transformation **T**, which occurs in the lens, we have:

$$Y = T\{X\} + \varepsilon$$

where the final image Y is impacted by the noise (ε) in the system and the transformation in the lens.

In design, **Y** can be the final design outcome. We are interested in **Y**, the 'solution.' For a design, if we take the inverse and transform this equation, as in the act of designing or managing a constructed ecology:

$$T(Y) = T^{-1}\{T(X)\} + T^{-1}\{\varepsilon\}$$
$$= X + T^{-1}(\varepsilon)$$

we can see that noise (ε) has a vital influence on the solution Y. Our consideration here is that Y equals an ecological design; thus, in:

$$Y = T\{X\} + \varepsilon$$

site X is transformed by the designer and is impacted upon by noise ε in the system.

Inverse problems raise two issues – first, incomplete data and, second, noise in the data. Inverse problems can deal with incomplete data; designers work continually with incomplete data or partial knowledge of a site – it is common for us to act without complete data or with only partial knowledge.[23] However, incomplete data, or a sense of incomplete data, can hold up decision-making in ecological science. Incomplete data can make scientists hesitate. Recently, Blanca Jiménez Cisneros noted that more data were needed before her group could devise policies to save wastewater.[24] The conservation community is also concerned that the sense of incomplete data leads to delays in conservation outcomes and delays in strategies to assist conservation.[25] The ecologist David Lindenmayer and his co-authors note that delays in decision-making due to incomplete data have contributed to species loss.[26] Yet in response to 'thinking backwards'[27] conservationists have taken the idea of the inverse problem and used it as a framework for insect conservation and for the articulation of outcomes in large-scale ecological networks for biodiversity retention.[28] Michael Samways and James Pryke noted that by adopting this retrospective analytical approach of thinking backwards, they were able to maintain biodiversity while still keeping productive forestry and landscape activities, such as grazing in southern Africa.[29]

Science is often caught in the trap of thinking that more data are better, often because our understanding about the world comes directly from data, and more has been better. However, science is now increasingly dealing with 'action-orientated outcomes' with which design also deals, where there is a combination of science, technology, and social-cultural issues.[30] These sorts of outcomes – for example, in conservation and restoration, issues with which designers also grapple – demand decisions within time frames that set deadlines on decision-making. Sometimes the time frames will be very clear, as in design deadlines, while others will be fuzzier, such as in conservation or 'greening'. But failure to reach the deadlines of conservation and restoration can lead to extinction and they need timely decision-making. Design has always had precise deadlines, and time pressure is an example of the increasing commonality between science and design.[31]

Another aspect of science-design commonality is the need for post-build performance testing of outcomes and dissemination of this information to other practitioners and scientists. For example, a landscape architect who had completed an award-winning storm-water wetland and frog habitat in Melbourne expressed frustration to me that there had been a lack of ecological assessment after the build. Did the frogs use the habitat? Our only evidence was going there with a student group, and we all heard frogs croaking on a damp day; however, were these the targeted rare species or just any frog? Ecologists and other scientists share the designer's concern for evidence-based practice.[32] In this case, more data through assessment post-build will assist future builds of frog habitat, underlying the need for post-testing that cannot be done by simulation, which I will discuss shortly.

Separating needed data from noise in the system

Noise is a distraction in data. Is it possible to identify noise in data? A marvellous and indeed frightening example of how someone needed to sort out noise from important data for the real final ambition was Qantas flight QF32 in November 2010 from Singapore to Sydney, when an engine exploded and debris fell on Indonesia; shrapnel degraded or destroyed important systems on the plane. Captain Richard de Crespigny wrote of this near-catastrophic engine failure in the Airbus A380, the world's largest passenger plane, carrying 440 passengers and 29 crew.[33] He later received the International Air Association's highest honours, and his handling of the plane that day was praised as 'the finest piece of airmanship in aviation history'.[34]

As a young trainee pilot some thirty-five years before, de Crespigny's first flying instructor had always admonished him to 'fly the plane'; that is, to fly the plane as an entity, not as a body of parts, and to focus on that as the sole essential aim of flying. In the QF32 crisis, the cockpit crew received continual deafening barrages of information from the plane's computer systems, relaying systems failures one by one. Taking each failed system as a piece of information, the computer suggested what should be done about that 'bit' of the plane. De Crespigny ignored them, knowing that many of warnings he was receiving were simply noise – data and detail that would not ultimately assist in the ambition of flying the plane. The data barrage was alarming, but was largely concerned with small separate issues that ignored interconnections between the vital systems of the plane. For two hours this incessant stream of noisy information beset the crew on the flight deck, but de Crespigny filtered out the important messages from distractions that worked against his sole ambition of keeping the plane in the air and capable of landing. The lesson from QF32 to design issues is that we also have to deal with an endless, perennial stream of data about all aspects of our site, and much of it is of little value to the decisions we need to make and the actions we need to take. This valid lesson from aviation reflects Leonard Bernstein's philosophy of teaching that the best way to know something is to place it in the context of another discipline.[35]

More data can be just noise, and not really assist us in our design ambition. Indeed, we will never know a system completely. This is an important point for those who work in data-poor environments, or where there is little assistance from allied professions, or where funding is too poor to allow more detailed data collection. Here, thinking backwards will occur despite major gaps in information.

There are two other important points to make concerning noise in ecological-designed systems. First, noise can occur in cascades[36] because interactions between factors can occur in cascades. 'Trophic cascades' is a term from ecology used primarily in connection with predator–prey relationships and the cascading impacts changes in predator–prey relationships have within a food web. However, the concept of trophic cascades is moving beyond experimental and theoretical ecology, and it can be used in the context of designed systems. Trophic cascades can be context-dependent and non-linear.[37] Second, some noise will work across multiple systems and perform with dynamic intersystem processes and networks,[38] as we would expect with ecological systems.

A central question for ecologists, mindful of noise in the data, is how to advise designers and how to comprehend and account for noise. In designs, noise might be heterogeneous, spatially patchy, not uniform temporally, and might differ markedly between sites, leading to different difficulties in sites[39] and different answers, even if about the same ambition. For example, if we were designing for rainwater harvesting, what might the noise be in design data? In a tropical country, noise will be quite different from that of a desert environment. Decisions and speculations about thinking backwards from a design ambition for a site can be made complex by data; there can be a sense of worry if data is 'incomplete', and noise in the data can distract us from really knowing the site, and from making decisions about that site. Our student working in Banda Aceh was being overwhelmed with data because, in fact, a lot of the data collected in such forward-problem approaches to studio were simply noise, and not needed. Some data can happily remain 'incomplete' because they will not greatly impact a design.

The tricky issue is what data to discard or do without. A potentially useful way of dealing with noise or incomplete data is for designers to filter categories of data to see more clearly the traps and snares of worries about noise and 'missing data'.[40] Filtering might liberate us from the awful sense of not having all the information about a site. For example, if we look at the issues behind suburban design (shown in Table 6.1), we might find that some data will be Robust and resilient to all major changes in the system (such as underlying strata and local ecological knowledge); some will act as a Fulcrum, being pivotal, and changeable (such as planning legislation, water availability, maintenance regimes); some information will be Scaffolding for other information (such as lot size, retention or destruction of topography, road widths, percentage of locally endemic species) – and these will impact others via cascades and provide the basis of other information or processes. Then, there is the dreaded Game-changer; the Game-changer might come out of the blue (such as a new freeway, new streetlights with increased glare, drought, war); some Game-changers may mask as Dormants before being felt; a Dormant will arise if new conditions occur (such as a new government, problems with groundwater, pest invasions, coming climate change). Data can also be filtered as a Dependent, which depends on other issues or conditions (dependents might be tree canopy, animal-species presence or absence, pedestrian access across a suburb). Independents are Uniques and operate in their own little knowledge tribes (such as street-light luminaire types); Externals[41] are out of local control (such as a sudden decision by a state government overturning a local council's planning policy from densities of two houses per lot to four). Outliers are possible Distractions, such as design solutions at sites socially and ecologically different to the one under consideration, and they can occur in design when precedents are considered – if the precedents are not considered carefully, they can lead us astray if they presume universal principles across geographic and ecological landscapes. Finally, I add a perhaps curious form of 'data category': Camouflaged. Camouflaged can be either friends or assassins; friends could be local knowledge that, while of assistance, may be repetitive or merely supportive of already well-known information; friends leading us astray might be those suggesting a default position, such as I outlined in Chapter 3 with the savannah theory. Assassins are likely to be Game-changers as well, and might be something as insurmountable as people who refuse to believe the data accumulated by years of research.[42]

Table 6.1 Data filtering for suburbs, showing potential data types, some of which will be noise and some vital for decision-making for the design of constructed ecologies within a suburb. All of these categories will change importance depending on location. Based on a shorter list in Grose, 2014.[43]

Categories of data	*Found in suburbia?*
Robust (resilient to change)	Underlying geomorphic conditions Local ecological knowledge
Fulcrum (pivotable, changeable)	Planning legislation changes Water availability Soil health Maintenance regimes
Scaffolding (impacts others via cascades)	Lot size Retention of topography Road widths Percentage of locally endemic plants
Game-changer	New freeway New streetlights with increased glare Drought
Dependents (on others)	Tree canopy Animal-species presence and absence Pedestrian access across the suburb
Independents (Uniques)	Streetlight types
Externals (out of local control)	Air pollution Climate and climate change Wind direction
Outliers (possibly loud but distractions)	Design precedents from climatically and culturally different regions
Dormants (arise if new conditions)	Problems with groundwater Pest invasions
Camouflaged (friends and assassins)	Histories held by elderly, long-time residents, indigenous communities Unmapped drainage lines Those who do not believe in data or cherish old unsupported assumptions

All of these possible types of data will carry noise and will be context-dependent. Thinking about types of data and potential noise does not solve any problem in a simple way, or give an immediate solution, but will help designers deal with unknown impacts by acknowledging them and stating the uncertainties in the design. It might also allow the articulation of sites that have particular suites of data and noise, and the mere identification of these will assist the design process. However, thinking backwards gives us a clear and constant objective.[44]

Ecological scientists are now facing similar uncertainties in modelling, with a view that 'incorporating rather than avoiding uncertainty will increase the chances of successfully achieving conservation and management goals'.[45]

Speculation in designing from the inverse problem

By designing towards an ambition, we usually know what that ambition is, or what a number of ambitions might be. Do we then have the answer when we start designing? In a sense we do because we know the target of our actions.

The uncertainties, various categories of data, senses of incomplete data, and issues of noise in the data – all work to promote speculation. Speculation can employ all the 'hows' and 'whats'[46] – all cultural, social, aesthetic, engineering, scientific, and health aspects of an ambition – as components making up the answer to the inverse problem. Speculation can be defined as the forming of a theory or conjecture without firm evidence,[47] or more broadly as the experimental modelling or representation of particular environments.[48] Science also speculates to form its theories and hypotheses. Many laymen do not know this.[49] Within design there is often a misunderstanding of speculation in science, and even a belief that speculation has no place in science, and that science is concerned with empirical method and problem-solving alone. This is not the case; science is inquiry, and there has been a long history of the role of speculation in natural science, 'natural philosophy', and Western philosophy as a whole.[50] Carl Sagan described science as 'a way of thinking much more than it is a body of knowledge'.[51] The central methodology of science is thinking.

Speculation in science is really a natural extension of the hypothesis, and for the last few centuries the hypothesis has been the solid workhorse of science. The definition of hypothesis is that 'it can be refuted or confirmed (to some degree) by observational or experimental means'.[52] Within these sound criteria, scientists speculate as a central part of their research process. In addition, speculation has been considered as a way of addressing verifiable ideas, but also 'alternative thoughts, whose presence is essential to give credibility to the hypothesis that is taken to be the truthful one'.[53] Speculation is also taken to be a necessary part of the interpretation of data.[54] Thomas Teo, who has written on this subject from the viewpoint of psychology, notes that the desire for objectivity might mean different things across the science–humanities divide: that objectivity in science might mean excluding subjectivity, while in the humanities objectivity might mean actively including subjectivity. This division might well be why designers can feel that science does not speculate. The point here is that science and design have more commonalities than often imagined by either group. Neither design nor science is one way of thinking, but mixes, and searches and pursuits to achieve a sound outcome whether built, conceptual, or data-driven. I feel that it is important to recognise and strengthen an understanding of this commonality because joining design and science more pointedly will not only be more fruitful but will be necessary in order to tackle major problems we are facing in the realm of constructed ecologies.[55]

Digital techniques and visualisation are assisting many disciplines of science in critical speculation. For example, computer-assisted visualisation has revolutionised the study of fossils, giving insights into what a now-extinct animal looked like as a living entity. In one example with a theropod dinosaur under study at the University of Bristol,[56] palaeontologists are interested in a functional analysis of the dinosaur's bite, way back in the Mongolian Cretaceous of 90 million years ago. They carried out Finite Element Analysis, which predicts a subject's responses to real-world forces such as heat, shock, vibration and fluid flow, and this allowed the testing of various skull configurations and feeding patterns. For the dinosaur under study, simulation allowed the reconstruction of possible feeding behaviour and insight into the development of the skull (Figure 6.2). The authors noted that 'each reconstruction step increases the degree of interpretation introduced ... particularly if performed manually'.[57] They worked with geometric morphometrics and automated reconstruction methods that could be assessed statistically and reproduced accurately. Their simulation of form and function in response to environmental conditions sounds like the territory of the designer of constructed ecologies.

Of great relevance to issues that designers face, members of this dinosaur research team noted that communicating the wealth of three-dimensional digital information, including interactive models and animations, is difficult in print publications.[58] To address the challenge of public engagement, they used three-dimensional PDFs, QR codes,

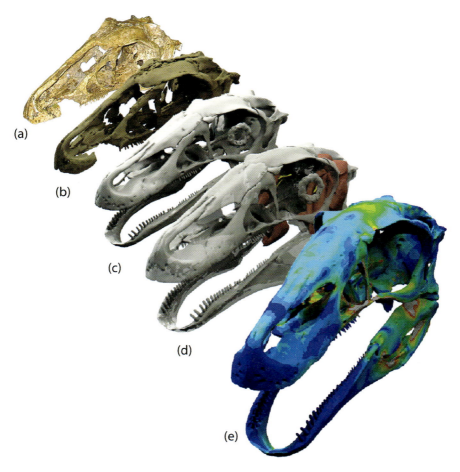

Figure 6.2 Step-by-step process, with each step requiring decision-making, was used to reconstruct this dinosaur. The images shows individual steps in the digital restoration and reconstruction process, exemplified by a model of the skull of *Erlikosaurus andrewsi*. (a) far left: Original fossil (lower jaws omitted due to disarticulation). (b) Digital representation of the fossil. (c) Restored cranial anatomy. (d) Restored skull with reconstructed jaw adductor muscles. (e) Final finite element model based on (c) and (d). Skull length is 260mm. Details from Cunningham, *et al.*, 2014.

Image: From Stephan Lautenschlager, with permission from Cell Press.

and anaglyph stereo-imaging – techniques all used in engineering and entertainment, as possible tools for outreach and explanation.

Joining precise scientific data to design in speculating about possible outcomes of particular design actions will be increasingly central to design. Landscape architects can achieve further speculation by examining feedback loops[59] and multiple-decision testing. Feedbacks and reiterations of design possibilities allow for testing of greater complexity in dynamic and changing systems. Digital techniques give opportunities for speculation to drive design change. In their recent book *Landscape Architecture and Digital Technologies*, landscape architects Jillian Walliss and Heike Rahmann have discussed specific digital techniques that can be (and have been) used as drivers of design in practice.[60] With new information on hand, real-time data can now be used to simulate and speculate on impacts of design changes, no matter how subtle. Working with data to move towards an ambition increases the scope by which imagination can experiment with more complexity and in dynamic, changing systems, examining feedback loops and testing 'what-ifs?'

One of the problems of two-dimensional mapping is that the data used emphasise representation and information 'determined from what is already known by the designer—whether analytical, creative or intuitive'.[61] In short, most design information springs from existing information given in all sorts of maps, books, planning documents; and details of physiography, history, ecologies, populations, and spaces. All of this information sits on the left-hand side of the equations given above as part of a forward problem. Data trapped in the forward problem's left-hand side are liberated by the ability of digital techniques to open up a dialogue between both sides of the equation. This dialogue also liberates the designer from the techniques and conceptual limitations of two-dimensional mapping techniques.[62] As Jillian Walliss and Heike Rahmann point out, dialogues provided by digital techniques enable a designer to actively test the performance of designs by setting up rules and limits under which a design might operate. Rules and limits give discipline to the design process, and liberate creativity by allowing the designer to command and test the impacts of design decisions. In turn, testing design decisions gives *iterations of possibilities* that can be further evaluated. All of this points, once again, to Lewis Mumfords' *power of restrictive conditions*. Restrictive conditions give us a number of things: exactness, precision, repeatability, principles, and rigour.

What are the impacts of these new ways of designing? First, the capacities liberated by digital tools take us well away from the old idea of 'digital as representation' to that of digital techniques as essays of inquiry with visual outputs. Second, the change to *iterations of possibilities* suggests that we are now in the realm of design as an experiment; this is quite different from the 'designed experiments' within any built design process. Science works with designed experiments very solidly (for all experiments are designed), but in landscape architecture designed experiments have been confined to building the idea, and then testing the design's success once it has been completed.[63] Ecologists and other scientists readily and appealingly understand such an approach of learning by doing, perhaps because it involves a method, action, and measurement of the outcome. However, this type of build or 'suck and see' experiment in design, and subsequent reflections on successes or failure, completely ignores the power of today's landscape architecture. The new digital 'toolkit' enables an expansion in design experimentation and creativity. Design as essays of inquiry through experiments, also the true realm of good science, relates to testing, speculation, and simulation of all sorts of possibilities using digital tools. Simulations can be modelled in the plastic form, a method particularly valuable and cost-saving in design, or through the use of actual physical models, a technique long carried out in environmental testing. As Jillian Walliss and Heike Rahmann noted, physical and digital simulation moves us beyond the conceptual limitations of techniques such as mapping.

Simulation appears to be a difficult thing to define, perhaps because it appears to lie between ideas and experimentation. We can readily ask: what is a simulation, and how does it relate to an experiment? In a stimulating paper, the philosopher Eric Winsberg outlined this complex relationship between simulation and experiment;[64] two points are of particular interest to design thinking. First, simulation has been considered as a 'technique that begins with well-established theoretical principles, and through a carefully crafted process, creates new descriptions of the systems governed by those principles'.[65] That is, by a certain type of experimentation, simulation provides 'information about systems for which previous experimental data is scarce' or even absent. From such information, a designer can make new inferences about the impacts of actions, or the impacts of design decisions, saving us a lot of time, heartache, and money. Second, we can see simulation as an entirely new mode of scientific activity sitting in the interesting space between theory and experimentation because simulation combines the techniques of experimentation and observation with theory and ideas. Simulation might cross disciplines more fluidly than in the typical experiment, where methodologies are often tied with disciplines.[66] In this way, simulation can provide a 'trading zone' between theory, techniques, and many disciplines. For designers, working with simulation conjointly with other disciplines provides access to information often unavailable to design practice.

Simulation using physical models can reveal behaviour that cannot be understood. A simulation test determined how a July 1958 landslide in Lituya Bay, in south-east Alaska, was able to generate a

Table 6.2 Factorial experiments in design allow more complex analyses from simple testing. The possibilities of what can be tested and considered in a factorial experiment go by the general formulae of Levels to the power of Factors (or independent variables) examined as a starting point. Here, there are 3 variables and 2 levels i.e. 2^3 (8), very simple combinations to test. They can be readily tested using digital techniques as design tools that have the capacity to examine both spatial and ecological impacts whilst in the plastic stage. Thus, we can join the facts to best fit by answering to the experience we have.

Variable or questions	Test 1	Test 2	Test 3	Test 4	Test 5	Test 6	Test 7	Test 8
Footpaths	wide	wide	wide	wide	narrow	narrow	narrow	narrow
Setbacks	large	large	less	less	large	large	less	less
Road width	wide	small	wide	small	wide	small	wide	small
Questions and considerations for each test	Visual implications (might be multiple) for each Test 1 to 8.							
	Implications and issues (ecological, planning, management) on other factors, inter-relationships, and transactions among components; transcending existing planning guidelines for each Test 1 to 8.							
	Ranking of factorial tests (poor, good, better, best) meeting intentions in thinking backward i.e., we can rank Tests 1 to 8, and articulate why.							
	Further thinking about best fit in context and in regard to the structures and variables in Tests 1 to 8.							

Conclusions

While mapping techniques remain essentially static land-use planning tools that attempt to identify critical aspects of a site, we can now change this stasis by testing and identifying critical processes within a design, both at the site and in experimental actions. As seen in Table 6.2, designs can be tested via digital iterations of design – 'What if we change a factor of the design?' – the 'what ifs?' of design ideas. Testing in this way allows the articulation of key components that might lead to successful projects or prevent success. Here, we can go well beyond mapping attempts to identify critical spatial aspects of a site and can test for processes among quite different sets of information and at various scales. Digital iterations allow us to determine how spaces work as systems. Designing in this way is a major change in operation and thinking from a static two-dimensional understanding. The same shift from static to active is seen in medical tomography (which uses the concept of inverse problems) in its move from spatial mapping to functional testing with functional magnetic resonance imaging (fMRI).[70]

The computational power of digital tools can bridge spatial and ecological processes and allow the testing of processes, design possibilities, and responses of design to changing scenarios. For example, in the restoration of the dryland river of Wadi Hanifa in Saudi Arabia[71] (see Figure 6.3), performative testing revealed how the river would respond to fluxing water levels in the wadi. The designers tested the spatial and ecological performance of the water purification systems put in place. Performative testing also examined cultural uses (socialising in family groups, cooling off, walking) and the response of local people to alternating opportunities for interaction along the designed site; here, the designers took into account particular religious and cultural sensitivities within the Islamic society (privacy), and social impacts of those within the site.[72]

Many educators in design still tend to see digital techniques as mere representation, largely due to the emphasis on the visual and not on

Figure 6.3 Wadi Hanifa, near Riyadh, Kingdom of Saudi Arabia, showing the restored water system of the wadi and its use as a cultural space. This system was tested digitally for the best opportunities to unite water, ecology, and community visits. This wadi is an important valley and natural water drainage, now being restored. The project won the Aga Khan Award for Architecture in 2010.

Image: Courtesy of Arriyadh Development Authority, http://www.burohappold.com/projects/project/wadi-hanifah-54/. Accessed 23 March 2016.

processes (whether ecological or cultural). However, digital techniques take us far beyond representation. A spectacular way in which digital techniques and massed data analysis have led to a new conceptual understanding and a new reality has been the work of graphic designers on the 2014 science fiction movie *Interstellar*. The movie screenplay required a 'wormhole' and a black hole. What do these look like, and what would they look like for the space traveller? Only a few years ago, this need would have been the realm of pure imagination and graphic cleverness. No longer. Kip Thorne, the world authority on gravity and what happens to gravity in black holes (it is distorted hugely), gave the thirty designers who worked on this issue for one year many algorithms concerning the performance of these stellar structures. The result is not only a visual feast, but is as accurate as we earthlings can be about the nature of wormholes and black holes. This film is a classic example of how computational power and testing of real data as restrictions and rigour in mathematics have arrived at a better understanding of how a system works, in this case the performance of a black hole.[73]

For a very different reason, Balmori Associates carried out a process of testing for the Beijing Garden Expo with the ambition of improving plant growth (Figure 6.4). In this garden, Balmori mapped areas of similar conditions and then constructed a parametric computational model of the garden that 'adapts to and aligns with transient information flows. Advanced programming methodologies allow the model to analyze year-round natural conditions of a particular area of the site, including sun hours per day, slope conditions, altitude, and wind exposure'.[74] Working with data and simulation, Jason Toh and Jillian Walliss worked on the best performative solutions in tropical Singapore, with the ambition of reducing urban heat and humidity and improving the thermal comfort of a major city street; they did this by detailed analysis of trees, paving, and street, and used simulation and parametric modelling to suggest a new type of street planting.[75] Importantly, and referring back to Chapter 2 in this book, their performative research work reveals the significance and power of working with global differences – what might work in New York might not work in Singapore – and warns against generic mediation.

In the Preface to this book, I referred to James Corner's comments concerning 'theory to elaborate rules and procedures for production',[76] that have tyrannised contemporary theory in landscape architecture. Mapping has been a limiting procedure in design. In contrast, computational methods employ a range of techniques and are not one 'method'. They are expressions of an inquiry, a search for solutions to a question posed by the site, and an inverse problem where we can imagine and dream of an ambition that is itself an 'answer'. Importantly,

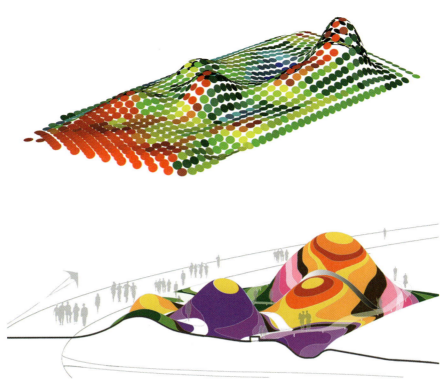

Figure 6.4 Sound Waves. In these examples of an examination through data collection of sun performance over time and space, the designers Balmori Associates were able to test the performance of plant growth on their site while in plastic form. Performative testing allows for a fine-grained assessment and response to real data.

Images: With permission from Balmori Associates, Landscape and Urban Design.

our testing can search backwards and forwards across spatial dimensions and time, and is not limited to a linear or forward manner.

Non-linear and multidirectional behaviours assisted human imagination as an exploratory process throughout human history. Working in southern Africa, the anthropologist Marlize Lombard drew a rugged fitness landscape and mountaineering analogy to explain human cultural, behavioural, and technological evolution and flexibility (Figure 6.5).[77] She considers that we had a flexible, fluxing, and evolving manner of working; we tested processes and worked with uncertainty throughout history. Her depiction of human testing is far from the linear and forward direction often assumed in design, and places current digital testing and experimentation as part of the long continuum of working with data.

Digital computational tools can test spatial and ecological processes, can respond to the local and are 'grounded in site', can accommodate real-time processes, and respond to physical, ecological, and social

Figure 6.5 In this mountaineering analogy for human cultural evolution, anthropologist Marlize Lombard reveals a little of the myriad ways there are to move and change in response to shifting and new data. This ties in with the discussion in Chapter 3 of *shifting adaptabilities*, and the discussion in this current chapter about the need for greater experimentation in landscape architecture.

Image: With permission from Springer.

data and test their changing impacts with various scenarios. The design power of these tools is taking us well away from overlay and spatial analysis and into the complexity of three-dimensional systems. Digital tools can monitor performances of proposed and completed designs, and give feedback that enables us to learn from the processes and modify new designs in response to greater knowledge. The union of science and design to achieve an ecological ambition creates ideas that are solid, tested, flexible, and able to respond to changing conditions over time; to deal with uncertainty and future needs; and to be tuned to sites that are different in scale, culture, ecological parameters and demands. Testing and performance are enabling designers to take part in the most pressing, difficult, and complex global issues, as well as giving us remarkable imagined ways of seeing, and revealing them to our scientific colleagues. Dealing with vast amounts of data from sensors, electronics, and networking systems will be a challenge and likely a delight to many designers negotiating the bridges between ecological science and design for constructed ecologies.[78,79]

A major impact of changes outlined in this chapter suggests that landscape architects should not be seen as doing a bit of everything, as has long been stated as a feature of the profession. Instead, young landscape architects might now consider specialising in particular fields, such as constructed wetlands, small-system water issues, sanitation, play spaces, green infrastructure, *design georgics*, forestry, urban heat, refugee camps, and Smart Villages – to name a few. Many specialisations will demand stronger links between design and science. Many (but not all) will need to consider multiple, tested solutions with an understanding of noise in these systems, whether cultural or ecological, and work with uncertain and variable restrictive conditions to achieve complex and multifaceted ambitions. In this way, the profession will have the intellectual power, knowledge, and agility to make solid and lasting contributions to major constructed ecologies and global social and ecological issues. That is our challenge.

Notes

1 This was taken up from Ovid's comments in *Ars amatoria* (AD 2) on 'how to get women'; clearly there are many ways. Quote is from: W.M. Calder, 1975, 'Ulrich von Wilamowitz-Moellendorff to Wolfgang Schadewaldt on the classic', *Greek, Roman and Byzantine Studies*, 16(4), 451–467. Available online at http://search.proquest.com.ezp.lib.unimelb.edu.au/docview/1301490579?accountid=12372.
2 D'Alembert, cited in J.G.A. Pocock, 2004, *Barbarism and Religion, Vol.1: The Enlightenment of Edward Gibbon*, Cambridge: Cambridge University Press, p. 188.
3 Kees Dorst, 2011, 'The core of "design thinking" and its application', *Design Studies* 32: 521–532, discusses framing as a term commonly used in design, and as if fundamentally different from reasoning in other fields such as science. This seems a simplification of science research practice.

4   Bryan Lawson and Kees Dorst, 2009, *Design Expertise*, Oxford: Architectural Press, p. 241.
5   Ian McHarg, 1969, *Design with Nature*, Garden City, NY: Natural History Press.
6   Tim P. German and H. Clark Barrett, 2005, 'Functional fixedness in a technologically sparse culture', *Psychological Science* 16(1): 1–5.
7   Ibid.
8   Edwin Abbott, 1884, *Flatland: A Romance of Many Dimensions*, London: Seeley and Co. Several new editions have been published, and it is also an eBook.
9   McHarg, 1969, p. 197.
10 Ken Robinson, 2001, *Out of Our Minds: Learning to be Creative*, Chichester: Capstone Publishing, p. 269.
11 I outline this in Margaret J. Grose, 2014, 'Thinking backwards can inform concerns about "incomplete" data', *Trends in Ecology and Evolution* 29: 546–547; and 2015, 'Inverse problem solving helps us collect the needed data: A reply to Falcy', *Trends in Ecology and Evolution* 30: 295–296.
12 J.B. Robinson originally used this term for energy policy. See Robinson, 1982, 'Energy backcasting: A proposed method of policy analysis', *Energy Policy* 10: 337–344.
13 Adrian D. Manning, David B. Lindenmayer and Joern Fischer, 2008, 'Stretch goals and backcasting: Approaches for overcoming barriers to large-scale ecological restoration', *Restoration Ecology* 14: 487–492.
14 Ibid., p. 490.
15 Elasticity analysis is taken from: Tim G. Benton and Alastair Grant, 1999, 'Elasticity analysis as an important tool in evolutionary and population ecology', *Trends in Ecology and Evolution* 14: 467–471.
16 Jacques Hadamard, 1902, 'Sur les problèmes aux dérivées partielles et leur signification physique', *Princeton University Bulletin*, pp. 49–52.
17 The defining book is Alfred North Whitehead, 1929, *Process and Reality: An Essay on Cosmology*, New York: Free Press.
18 Charles Sanders Peirce, 1878, 'How to make our ideas clear', *Popular Science Monthly* 12: 286–301. Available online at http://www.peirce.org/writings/p119.html/ Accessed 7 November 2015.
19 Robert Burch, 2014, 'Charles Sanders Peirce', in Edward N. Zalta (ed.), *The Stanford Encyclopedia of Philosophy*, at http://plato.stanford.edu/archives/win2014/entries/peirce/ Accessed 12 June 2016. Burch discusses in succinct terms Peirce's logic and ideas. Peirce had articulated three types of thinking about problems: induction, deduction, and abductive inference.
20 Kyle Daun, *Inverse analysis*, available online at http://www.kjdaun.uwaterloo.ca/research/inverse.html/ Accessed 12 June 2016.
21 Dorst, 2011, p. 524.
22 With thanks to the late Peter Hall FRS for this analogy.
23 Jeff Dozier and William B. Gail, 2009, 'The emerging science of environmental applications', in Tony Hey, Stewart Tansley, and Kristin Tolle (eds), *The Fourth Paradigm Microsoft Research: Data-Intensive Scientific Discovery*, Redmond, VA: Microsoft Research, pp. 13–19, available online at http://fourthparadigm.org/.
24 Blanca Jiménez Cisneros, 2013, 'Water management: The data gap', *Nature* 502: 633–634.
25 David Lindenmayer, Maxine P. Piggott, and Brendan A. Wintle, 2013, 'Counting the books while the library burns: Why conservation monitoring programs need a plan for action', *Frontiers in Ecology and Environment* 13: 549–555.
26 T.G. Martin, *et al.*, 2012, 'Acting fast helps avoid extinction', *Conservation Letters* 5: 274–280.

27 Based on: Margaret J. Grose, 2014, 'Thinking backwards can clarify concerns about "incomplete" data in ecology', *Trends in Ecology and Evolution* 29(10): 546–547; and Margaret J. Grose, 2015, 'Inverse problem-solving helps us collect the needed data: A reply to Falcy', *Trends in Ecology and Evolution* 30(6): 295–296.
28 Michael J. Samways, 2015, 'Future-proofing insect diversity', *Current Opinions in Insect Science* 12: 71–78.
29 Michael J. Samways and James S. Pryke, 2015, 'Large-scale ecological networks do work in an ecologically complex biodiversity hotspot', *Ambio* 45(2): 161–172.
30 Carolina Murcia, James Aronson, Gustavo H. Kattan, David Moreno-Mateos, Kingsley Dixon, and Daniel Simberloff, 2014, 'A critique of the "novel ecosystem" concept', *Trends in Ecology and Evolution* 29: 548–553. Discussed in Margaret J. Grose, 2015, 'Inverse problem-solving helps us to collect the needed data: A reply to Falcy', *Trends in Ecology and Evolution* 30: 295–296.
31 I discuss these in Grose, 2014, 'Gaps and futures in working between ecology and design for constructed ecologies', *Landscape and Urban Planning* 132: 69–78.
32 W.J. Sutherland, *et al.*, 2004, 'The need for evidence-based conservation', *Trends in Ecology and Evolution* 19: 305–308.
33 Richard de Crespigny, 2012, *QF32*, Sydney: Macmillan.
34 Carey Edwards, 2013, *Airmanship*, London: Blacker Ltd, p. 18.
35 Leonard Bernstein, 1973, *The Unanswered Questions: Six Talks at Harvard*, Cambridge, MA: Harvard University Press. This is also available as an Internet video.
36 Iddo I. Eliazar and Michael F. Shlesinger, 2013, 'Noise cascades and Lévy correlations', *Journal of Physics A: Mathematical and Theoretical* 46: 392001.
37 Michael L. Pace, Jonathan J. Cole, Stephen R. Carpenter, and James F. Kitchell, 1999, 'Trophic cascades revealed in diverse ecosystems', *Trends in Ecology and Evolution* 14(12): 483–488.
38 Eric de Silva and Michael P.H. Stumpf, 2005, 'Complex networks and simple models in biology', *Interface* 2(5): 419–430.
39 Peter Hall and J.L. Horowitz, 2005, 'Nonparametric methods for inference in the presence of instrumental variables', *Annals of Statistics* 33: 2904–2929.
40 Adapted from Margaret J. Grose, 2014, 'Thinking backwards can inform concerns about "incomplete" data', *Trends in Ecology and Evolution* 29(10): 546–547.
41 A strong external in conservation would be the price on ivory that impacts elephant poaching.
42 Assassins will no doubt inhibit ambitions to reduce artificial light at night in cities, as discussed in Chapter 5.
43 Margaret J. Grose, 2014. This paper discusses the use of data filtering to obtain clarity. See also: Margaret J. Grose, 2015, 'Inverse problem-solving helps us collect the needed data: A reply to Falcy', *Trends in Ecology and Evolution* 30(6): 295–296.
44 General Jim Mattis of the US Marines gives a similar scenario concerning the importance of clear and constant objectives, available online at https://www.youtube.com/watch?v=tKIJKQRb53o from the Hoover Institute. Accessed 7 April 2016.
45 Benjamin S. Halpern, Helen M. Regan, Hugh P. Possingham, and Michael A. McCarthy, 2006, 'Accounting for uncertainty in marine reserve design', *Ecology Letters* 9: 2–11.
46 Kees Dorst, 2015, *Frame Innovation: Create New Thinking by Design*, Cambridge, MA: MIT Press.
47 From the *Oxford English Dictionary*.

48 Daniel Stokols, 1993, *Strategies of Environmental Simulation: Theoretical, Methodological, and Policy Issues*, New York: Springer.
49 See John Wright Buckham, 1917, 'Speculation in science and philosophy', *The Open Court*, vol. 12, Article 2. Available at: http://opensiuc.lib.siu.edu/ocj/vol1917/iss12/2/.
50 Outlined in Thomas Teo, 2008, 'From speculation to epistemological violence in psychology: A critical-hermeneutic reconstruction', *Theory & Psychology* 18(1): 47–67.
51 Carl Sagan, 1979, 'Can we know the Universe?: Reflections on a grain of salt', in *Broca's Brain: Reflections on the Romance of Science*, New York: Random House, pp. 13–18, at p. 13.
52 M. Bunge, 1983, 'Speculation: Wild and sound', *New Ideas in Psychology* 1: 3–6. This article is discussed in Teo, 2008.
53 D. Bakan, 1975, 'Speculation in psychology', *Journal of Humanistic Psychology* 15: 17–25.
54 Teo, 2008.
55 I discuss these commonalities in Grose, 2014, 'Gaps and futures in working between ecology and design for constructed ecologies', *Landscape and Urban Planning* 132: 69–78.
56 John A. Cunningham, Imran Rahman, Stephan Lautenschlager, Emily J. Rayfield, and Philip C.J. Donoghue, 2014, 'A virtual world of paleontology', *Trends in Ecology and Evolution* 29(6): 347–357.
57 Ibid.
58 Stephan Lautenschlager and Martin Rücklin, 2014, 'Beyond the print – virtual paleontology in science publishing, outreach, and education', *Journal of Paleontology* 88(4): 727–734. Stephan Lautenschlager has some delightful animations and images on his website, available online at http://eis.bris.ac.uk/~glzsl/functional_morphology.html/. Accessed 7 November 2015.
59 As noted by Charles Peirce.
60 Jillian Walliss and Heike Rahmann, 2016, *Landscape Architecture and Digital Technologies: Re-Conceptualising Design and Making*, Abingdon: Routledge.
61 Jillian Walliss and Heike Rahmann, 2016, 'The Experimental Nature of Simulation', *LA+ Interdisciplinary Journal of Landscape Architecture: Simulation*: 40–45. 'In a significant change, processes of simulation model knowledge rather than information, conceived as an experiment requiring the identification of specific functions, parameters and conditions. Considered within the context of landscape architecture, simulations present the testing of an epistemological understanding of nature which generates rules and limits which can inform new modes of designing and planning.'
62 Ibid.
63 Such as in Alexander Felson, 2005, 'Designed experiments: New approaches to studying urban ecosystems', *Frontiers in Ecology and the Environment* 3(10): 549–556.
64 Eric Winsberg, 2003, 'Simulated experiments: Methodology for a virtual world', *Philosophy of Science* 70(1): 105–125.
65 Ibid., p. 116.
66 Margaret J. Grose, 2010, 'Small decisions in suburban open spaces: Ecological perspectives from a Hotspot of global biodiversity concerning knowledge flows between disciplinary territories', *Landscape Research* 35(1): 47–62.
67 The landslide was caused by an earthquake. A local man and his 8-year-old son, who were in their boat in the bay, survived, the boat riding above the trees.

68 See 1958 Lituya Bay Megatsunami Simulation, BBC, available online at https://www.youtube.com/watch?v=yqbSsHp2q54 for the images of this simulation.
69 Cultural factors, such as rangeland management, traditional farming or herding practices, and community and historic aspirations for sites, can all be tested digitally.
70 R.B. Buxton, 2013, 'The physics of functional magnetic resonance imaging (fMRI)', *Report Progress Physics* 76 096601. Available online at doi:10.1088/0034-4885/76/9/096601.
71 Sareh Moosavi, Margaret J. Grose, and Jillian Walliss, 2015, *Performative Design in Restoration of Dryland Rivers*, International Federation of Landscape Architecture Conference, St Petersburg, Proceedings of the 52nd International Federation of Landscape Architects, World Congress, pp. 616–622. Available online at https://www.researchgate.net/publication/282691743_Performative_Design_in_Restoration_of_Dryland_Rivers/ Accessed 7 October 2016. A wadi is a bed or channel of a stream in south-western Asia and northern Africa that is usually dry, except in the rainy season.
72 Sareh Moosavi, Jala Makhzoumi, and Margaret Grose, 2015, 'Landscape practice in the Middle East: Between local and global aspirations', *Landscape Research* 41(3): 265–278.
73 For images concerning this see Kip Thorne's diagram of how a black hole distorts light, in Kip Thorne, 2015, *The Science of Interstellar*, New York: W.W. Norton, available online at http://www.wired.com/2014/10/astrophysics-interstellar-black-hole/ Accessed 5 July 2015.
74 Balmori Associates website, available online at http://www.balmori.com/portfolio/sound-waves?rq=Beijing.
75 See Jason Toh and Jillian Walliss, 2016, 'Design strategies for modifying the experience of humidity in Singapore's Orchard Road', paper presented at the 4th International Conference on Countermeasures to Urban Heat Island, at the National University of Singapore. Available online at: https://www.researchgate.net/publication/303824682_Design_Strategies_for_Modifying_the_Experience_of_Humidity_in_Singapore's_Orchard_Road.
76 James Corner, 2014, 'Three tyrannies of contemporary theory', in *The Landscape Imagination: Collected Essays of James Corner 1990–2010*, New York: Princeton Architectural Press, p. 87.
77 Marlize Lombard, 2016, 'Mountaineering or ratcheting? Stone age hunting weapons as proxy for the evolution of human technological, behavioral and cognitive flexibility', in M.N. Haidle, N.J. Conard, and M. Bolus (eds), *The Nature of Culture*, Dordrecht: Springer, pp. 135–146.
78 John H. Porter, Paul C. Hanson, and Chau-Chin Lin, 2012, 'Staying afloat in the sensor data deluge', *Trends in Ecology and Evolution* 27(2): 121–129.
79 This problem of data has also been identified for new farming initiatives, where farmers in many developed countries are concerned about the data deluge, discussed in Chapter 4 as an opportunity for landscape architects.

CONCLUDING COMMENTS

This is unabashedly a theoretical book. I have been selective in these chapters in order to address pressing concerns that I believe will enliven the profession of landscape architecture towards working more rigorously with contemporary issues. In doing so, I have encompassed a great range of material and ideas for current design and the future of landscape architecture, and I have taken discussion beyond the walls within which most discussion currently sits.

A key question permeating this book has been how landscape architecture might embrace knowledge from ecological science and other disciplines. If we do not meet this question head on, landscape architecture will flounder amid the burgeoning suites of new information, and collaborators from other disciplines will not look to landscape architects to create new projects. To assist us to make better designs I have drawn directly from science as an ally in two ways – by information that assists ideas of performance, and by the wealth of inquiry and theoretical ideas that designers can imaginatively morph and build upon. Science assists us to understand processes, metabolisms, systems, changing histories of landscapes and our place in them, behaviours of phenomena, and methods of controlled decision-making. Scientific knowledge imposes a rigour and vigour on our design process by the very limitations and boundaries it sets, as Lewis Mumford noted nearly fifty years ago. Rigour in our ambitions in design will renew, rejuvenate, challenge, and overturn old ideas and assumptions about the built environment.

It is my hope that those in the design profession, particularly those who are embarking on the journey into landscape design, will have found something here to inspire, excite, or intrigue them and, in doing so, will be able to use science more rigorously than often found in much landscape discourse, even that which purports to use science or its methods. Ecological science has moved enormously in the last few decades to embrace biotic instability, flux, diversity, disequilibrium, open systems, nonlinear pathways and relationships, human desire and perception, and shifts in responses with time, holism, and harmony between biotic and natural abiotic processes.[1] All of these offer great challenges to designing for the future and speak strongly to new approaches in landscape architecture.

In these essays my focus has been to boost the role of ecological science in decision-making and knowledge transmission, the legacies of differences in regions in *spectrums of responses* and *shifting continuities*, the articulation of climate spaces and the exploration of heterogeneous landscapes; the power and capacity of human adaptation to change and the dynamic flexibility within us in *shifting adaptabilities*; to work towards new possibilities of design engagement; to think through issues for multiple voices and not the ready ease of single issues (that are so often funded) leading to knowing the world in pieces; 'the things of the farmer' – *design georgics* – will expand the discipline, as we focus manners of working with strong inquiry and not assumptions. We may start to think differently and specifically backwards – to adapt the fitness of design to variation and possibilities, to work with the ideas of performance, testing, simulation, experimentation, and change, and to use the full *power of restrictive conditions* and uncertainty. These suggestions expand the horizons of landscape architecture.

Core to the challenges ahead for designers, as for ecologists, is the ability to engage communities that have less knowledge and fewer and shorter contacts with the natural world than any generation in human history. Additionally, we need to deal with a world population increasing due to the rubric of equating growth with progress or to religious observances, despite critical pressures on the environment that will lead to political instability.

I began this book with landscape as a historical subject, fluxing and changing. As our Mother the Earth is once again changing with time, one challenge will be to free ourselves from describing problems or working with wide generalisations under a sustainability umbrella that often shelters us from the real rain and the real storm, and to take the future as a gift and not as a threat. Rather, we must embrace the detailed difficulties ahead as questions and challenges posed for us to meet through design and science with our full creativity, with many niches for future design students and rich engagement with engineering and the sciences. In this book I have hoped to point to a more positive future for which we must work with all the knowledge and inquiry at our disposal, wherever that critical knowledge might lie.

Note

1 For a discussion of this view of science, inspired by soils, see: Henry Lin, 2014, 'A new worldview of soils', *Soil Science Society of America Journal* 78: 1831–1844. Fittingly, the last footnote in this book cites soils, where I began this journey. It might reflect the thinking of the soil scientist and the melding of both biotic and abiotic processes in that three-dimensional space.

INDEX

Abbott, E. 158
accounting 99
action-orientated outcomes 164
Adams, R. 138
adaptability *see* shifting adaptabilities
Africa 19
agricultural revolution: (medieval) 100; (neolithic) 98
agriculture 96–7, 117–18; biological conservation 104–5; climate years 106–7; data 111–13; definition 98–102; design/designers 108–14; non-crop species diversity 107; regenerative 109–11; soils 104; urban agriculture 96–7, 113–16; voices 102–3
Amazonia 99–100
Amidon, J. 112
ancient languages 77, 79
Anthropocene 5–6
anthropocentrism 5
anthropology 60, 68–9
Ardipithecus ramidus 68–9
artificial lighting *see* light/lighting
assisted migration/colonisation 33–4
associations 32–3
Australia 18–19, 79–80, 96, 131–2

backcasting 160 *see also* inverse problems
Balmori Associates 177–8
Barker, G. 98
beech (*Fagus sylvatica*) 21
behavioural flexibility 79
Bergson, H. 162

Bernstein, L. 165
Big Data *see* data
biodiversity 26, 40–1; agriculture 101, 104–8, 111
biome 15
boreal forest regions 28
brain, and plasticity 74–6

Canada 28
cancer 135–6
carbon farming 111
cave art 57, 77–8, 125
cereals, domestication 99
change *see* shifting adaptabilities; shifting continuities
chimpanzee behaviour 69–70
China 116–17
chloroplast DNA 23
Christensen, B. 65
circadian system 131–3, 135–6
Cisneros, B. J. 163
civic ecology 115
Clark, B. 133
Clement, G. 107
climate change 6, 30, 71, 101
climate space 31–3, 44; agriculture 106–9; connectivity 34–5; long-tailing 38–40; migration routes 33–4; refugia/stepping stones/holdouts 35–6; topographic buffering 36–8
climate variability 71
climate years 106–7
climate zone 15
Clyne, D. 42

INDEX

co-evolution 60
cognitive 81
cognitive archaeology 73
colour 42–3; horses 77; spectrum 129–30
computational methods 169–73, 175–80
connectivity 34–5
constructed ecology 7–10
continuities *see* shifting continuities
Corner, J. xiii, 4, 41–3, 177
corridors 34–5
counterpoint 96
crime 136–8
crop diversity 106–7
cultural construct 4
cultural evolution 74–5

d'Alembert, J. d. R. 155–6
Danby, R. 29
Dark-Sky Parks 144
darkness 127–8, 144
Dart, R. 60
data 111–13, 155–6, 159–60; categories 166–7; incomplete data 163, 166; noise 164–8
Dawkins, R. 41
daylighting rivers 161
de Crespigny, R. 164–5
decision-support 111
Denisovans 56, 67–8
design georgics 96, 108–14, 118, 186
design optimisation 162
designed experiments 171–2
Desvigne, M. 27–8
Diderot, D. 5
digital techniques 169–73, 175–80
dimmers/dimming 140, 142–4
dispersal *see* plant movement
distribution 30
DNA 23; genetic plasticity 73–5
domestication 99
Dorst, K. 157, 162
Dramstad, W. 34

ecological science 185–6
ecosystem services 5
elasticity analysis 160

embedded lighting 141–2
empiricism 6–7
energy 138–9
English landscape 58
environment 3–5
environmental generational amnesia 74
ethics 34
Europe 18, 20, 26
eurytopic living 70–1
exoticness 40–3
eye 133–4

factorial experiment 174–5
farms/farmers *see* agriculture
feedback loops 171, 174
Fens (England) 8–10
filtering (data) 166–7
flexibility (behavioural) 79
food production *see* agriculture
Forman, R. 34
forward problem 159–60, 171, 173
fossils 67
functional fixedness 157–8

Garden Island 131–2
genetics 23, 73–5
georgics *see* design georgics
Germany 138
Gibbs, J. 109, 111
Gioia, P. 42
glaciation 17–19
glare 137–8
Goucher, C. 98
grass 64–5
grass-bank farming 109
green corridors 34–5
green revolution 100–1
groundedness 43

Hannah, L. 36
health: LED lighting 140–1; melatonin 129, 131–3, 135–6
Heerwagen, J. 54, 58, 64
hemispheric differences 17–19, 26, 42; northern purity, southern richness 26–7
Hewitt, G. 20, 26

INDEX

holdouts 35–6
Holocene 6
Homo habilis 65–6
Homo sapiens 55–7, 65–8; light 132–6
homogenisation 40–1
Hopper, S. 42
horticulture 115–16
Hulme, M. 6, 71
human evolution 59–60, 65–8, 179; brain plasticity 74–6; landscape cognitive impact 72–4; landscape type 68–72; thinking 77–80
humans *see Homo sapiens*
Humboldt, A. von 15–16
Hunter, M. C. 34
hypothesis 168–9 *see also* speculation

Ice Age 18–19; agriculture 98–9; dispersal 20–5, 28; refuge 19–20
identity 77, 79
incomplete data 163, 166
insects 131–2
International Dark-Sky Association (IDA) 144
Interstellar 177
intervention 33–4
inverse problems 160–4, 173–4; noise 165–8
iterations of possibilities 171

Jeremiah, E. 79
Jevons' Paradox 138

knowledge 4, 7 *see also* scientific method

Lamarckism 73–4
land-sharing 108–9
land-sparing 109
Landman, K. 115, 117–18
landraces 107
language 77, 79
Lawler, J. 33
Lawson, B. 157
Le Nôtre, A. 3–4
Leakey family 65–7
Leakey, M. 67
LED lights 137–41

light pollution 128–9
light/lighting 126–8, 144–5; assumptions 138–9; ecological aspects 130–2, 141; eye 133–4; LED lighting 139–41; light pollution 128–9; melatonin/circadian system 132–3, 135–6; safety 136–8; spectrum 129–30; urban design 141–4
Lindenmayer, D. 163
Lombard, M. 179
long-tailing 38–40
Longcore, T. 128
Lucretius 31

McHarg, I. 43, 157, 159
macro-ecological history 17
Malafouris, L. 73, 75
managed relocation 33–4
mapping techniques 158–9, 171, 177
Maslin, M. 65
Mason, R. 19
Material Engagement theory 73
melatonin 129, 131–3, 135–6
metaplasticity (brain) 75
Meyer, E. 43
micro-refugia 35–6
migration (assisted) 33–4
Mizon, B. 137
modelling *see* digital techniques; simulation
Mohan, J. 28
monocultures 101
Mount Kosciuszko 19
mountains 18–19
Mumford, L. 43, 111, 171, 185

national parks 30, 117
nativeness 40–3
natural capital 5
nature 3–5; objectification 79–80
Neanderthals 53–4, 56–7, 59, 67–8
nest-building 69–70
neurology 74
New Zealand 19
night lighting *see* light/lighting
night sky 125–6, 128, 144–5
noise (data) 164–8

normal problems 161
North America 18, 26, 28
Northern Hemisphere 18–19, 26, 28, 42; northern purity, southern richness 26–7
Norton, D. 104
novel ecosystems 21, 31

objectification (nature) 79–80
objectivity 41, 169
Olden, J. 33
Olduvai Gorge 66–7
Olson, J. 34
Orians, G. 54, 57–9, 64–5
Orongo Station 109–10
overlay technique 157–9 *see also* two-dimensional techniques

palaeoecology 65
pathways 142
Peirce, C. S. 162
philology 77, 79
photographic representation 79–80
plant movement 15; hemispheric differences 17–19, 26; landscape planting 26–8; post Ice Age dispersal 20–5; projected futures 28–9; refugia 19–20
Pleiades 125–6
post-build performance 164
Potts, R. 71
precision agriculture 112
problem-description phase 159
protected land *see* national parks
Pryke, J. 163

Rackham, O. 5
Rahmann, H. 171–2
Reed, C. 115–16
refugia 19–20, 35
regenerative agriculture 109–11
Reid, N. 104
relativism 6–7, 41
Renfrew, C. 77
restrictive conditions 44, 111, 171, 173, 185–6
Rich, C. 128
risk 136–8

roads 143
Robinson, K. 160
Roosengaarde, D. 142
rural studio 97, 102
Russia 28

safety 136–8
Sagan, C. 168
Samways, M. 163
savannah 61–5
savannah theory 54, 57–60, 80–1; brain plasticity 74–6; defining savannah 61–5; hominin mind 77–80; human evolution 65–8; landscape cognitive impact 72–4; landscape type 68–72
scientific method 7, 156, 159, 164, 174, 185; speculation 168–73
second nature 107
seeds *see* crop diversity
self 77, 79
Serengeti Plains 62, 64
shifting adaptabilities 71–2, 80–1, 100
shifting continuities 17, 29–31, 44, 186
shifting places 32
Silvertown, J. 5
simulation 172–3, 177 *see also* digital techniques
site analysis 155–7, 159; overlay technique 157–9
skyglow 129
sleep *see* melatonin
Smart Highways 143
Smart Villages 101
Snow, C. P. 7
soft inheritance 73–4
soils 104
South America 19
Southern Hemisphere 18–19, 27, 42; northern purity, southern richness 26–7
specialisation 180
spectrum (colour) 129–30
spectrum of importance 43
spectrums of responses 17, 44, 186
speculation 168–73
Sphinx 14

spruce 28–9
Starpath 142
stars *see* night sky
stepping stones 35–6, 38
Stevens, R. 136
Stewart, C. 96
Stewart, F. 70
stress 75, 98
stretch goals 160
subject/object 79–80
Sumer 99

Teo, T. 169
Thomson, W. 129
Thorbeck, D. 97
Thorne, K. 177
Toh, J. 177
toolmaking 65–6
topographic buffering 36–8
Toynbee, A. 71
trees 21–5, 28–9; artificial light 130–1; savannah 61
trophic cascades 165
two-dimensional techniques 158–9, 171, 177

UK 137, 142
Uluru 79
uncertainty 168, 179–80
urban agriculture 96–7, 113–16
urban design (light/lighting) 141–5

variability *see* climate variability
visualisation 169–70

Wadi Hanifa 176
Walliss, J. 171–2, 177
Weller, R. xiii, xiv, 30, 96
Wexler, B. 74, 76
White, F. 61
white oak (*Quercus*) 21, 23–4
Whitehead, A. N. 162
Winsberg, E. 172
Wolff, E. 6
Woltz, N. B. 109–10
Woltz, T. 109, 117
wood-pasture hypothesis 72
Worster, D. 108
writing 99

Zimmermann, M. 3

Taylor & Francis eBooks

Helping you to choose the right eBooks for your Library

Add Routledge titles to your library's digital collection today. Taylor and Francis ebooks contains over 50,000 titles in the Humanities, Social Sciences, Behavioural Sciences, Built Environment and Law.

Choose from a range of subject packages or create your own!

Benefits for you
- Free MARC records
- COUNTER-compliant usage statistics
- Flexible purchase and pricing options
- All titles DRM-free.

Free Trials Available
We offer free trials to qualifying academic, corporate and government customers.

Benefits for your user
- Off-site, anytime access via Athens or referring URL
- Print or copy pages or chapters
- Full content search
- Bookmark, highlight and annotate text
- Access to thousands of pages of quality research at the click of a button.

eCollections – Choose from over 30 subject eCollections, including:

Archaeology	Language Learning
Architecture	Law
Asian Studies	Literature
Business & Management	Media & Communication
Classical Studies	Middle East Studies
Construction	Music
Creative & Media Arts	Philosophy
Criminology & Criminal Justice	Planning
Economics	Politics
Education	Psychology & Mental Health
Energy	Religion
Engineering	Security
English Language & Linguistics	Social Work
Environment & Sustainability	Sociology
Geography	Sport
Health Studies	Theatre & Performance
History	Tourism, Hospitality & Events

For more information, pricing enquiries or to order a free trial, please contact your local sales team:
www.tandfebooks.com/page/sales

 The home of Routledge books

www.tandfebooks.com